CHALLENGING THE UNKNOWN

CHALLENGING THE UNKNOWN

ALFREDO E. PHIPPS, JR.

Alfredo E. Phipps, Jr.

Copyright © 2020 by Alfredo E. Phipps, Jr.

All rights reserved. No part of this book may be reproduced in any manner whatsoever without written permission except in the case of brief quotations embodied in critical articles and reviews.

ISBN: 978-1-7358007-2-1 / E-book: ISBN: 978-1-7358007-3-8

Library of Congress Control Number: 2020920832

This book pays tribute to all those who supported me and who became an inspiration to me in different ways. You helped me design the characters in this book. Each character is a life experience that we all go through.

To my beloved wife, Yamiri: Your unconditional love sustains me. To my son, Joshua G. Phipps: I am proud of you. To my stepchildren, Michael and Mikela Gordon, my brothers and sisters Aristides, Samuel, Yendi and David, Miguel and Jenifer; and their respective partners.

Epigraph

"Purpose is an essential element for you. It is the reason you are in this world at this particular time in history. Your very existence is wrapped up in the things you are here to accomplish. Whatever you choose for a career, remember, the struggles along the way are only meant to shape you for your purpose."

-Chadwick Boseman
(1976-2020)

(A percentage of the sales will be donated to organizations that help people with cancer.)

Disclaimer

This novel is entirely a work of fiction. The names, characters and incidents presented in it are the work of the author's imagination. Any resemblance to real people, living or dead, events or locations is entirely coincidental.

Alfredo E. Phipps, Jr. asserts the moral right to be identified as the author of this work.

Alfredo E. Phipps, Jr. is not responsible for the persistence or accuracy of the URLs of external or third party websites referred to in this publication and does not guarantee that the content of such websites is, or will remain, accurate or appropriate.

Designations used by companies to distinguish their products are often claimed as trademarks. All brand and product names used in this book and on its cover are trade names, service marks, trademarks, and registered trademarks of their respective owners. The publishers and the book are not associated with any product or vendor mentioned in this book. None of the companies referred to in the book have endorsed the book.

Contents

Dedication   v
Epigraph   vi
Disclaimer   viii
Prologue   xi

1. It's Not Always About Winning   1
2. In the Parallel World   10
3. Too Difficult to Start Over   24
4. In the Big City, All by Yourself   44
5. Turning the Pages   62
6. The Lovers Reunite   80
7. Opportunities   96
8. The "Friday" Affair   116
9. Proposals   147
10. My Land's Calling   160
11. The Guarantor   167
12. Another Chance   176

End Note 183
Author's Commentary 185

Prologue

Didn't it occur to you ever in your life that sometimes, we need to adopt change; sometimes we need to go in the flow?

Sometimes, we have to let things be and wait for them to turn out right.

However, on the other side, some circumstances call for action; they call for attention and a prompt measure.

It is up to us to decide the route we would like for ourselves.

What do we choose? An initiative or a regret, about not taking the initiative when it was due!

It is often said only dead fishes go with the flow; however, is it even the case?

Can we make things turn just like we want?

I think not!

Many things happen in our lives that are out of control; there are instances where we fail to accomplish our goals or where things do not go as per our plans.

Should it make us accept our failure?

Enclosed are stories of people coming from various backgrounds, nationalities, and age groups, from a young boy to an old lady. All of them have been met with unexpected adversities in life.

However, should these adversities stop them in any way? Should they give in to adversities?

At times, they, like you, felt that they could not go forward, that they need to give up or it is too much to bear.

However, it is not always the case. Coal under pressure becomes a diamond.

This is what life expects from the characters in this book and from all of us.

Life puts different things in front of us. It is our approach to such events that shape our lives. No doubt, our experience helps us to make decisions. But life is a teacher.

This teacher teaches us lessons.

We all are given a different type of exam. We are put through different kind of situations.

It is not the case that God loves us less and loves someone else more.

We are all given challenges based on our capacities. We are all born and raised in societies different from one another.

We all have different personalities and different outlooks to life.

Every character in this book is faced with a different challenge.

It is not sufficient to say that one has a more difficult battle than others.

How will they come about? This is yet to be seen ...

I

It's Not Always About Winning

"Let's give a round of applause for the Jefferson Jaguars, the Roosevelt Rockets, and their incredible basketball players. On the count of three: one, two, three!"

The crowd clapped and cheered for the two teams as they prepared to enter the basketball court.

Everyone could hear an enthusiastic audience and their hustle and bustle from Jefferson High School's basketball court.

Packed with students, parents, and people from the community—all basketball fans—the crowd knew this game would be the one to watch.

The local police were on high alert and kept a close watch on the nearby streets to ensure everyone attending the semifinals could enjoy the game.

It was a proud moment in the school history, as this was the first time that Jefferson High School had made it to the state basketball semifinals.

Among the crowd were two proud parents, Peter and Jessica Johnson.

It was because of their son Ariel, the star point guard of the Jaguars, that the team made it to the semifinals.

Although the team had lost a few points during the third quarter, they soon found their rhythm and they were only trailing the Rockets by one point.

Peter and Jessica felt confident that their son could win the game in the final minutes of the fourth quarter.

It was now or never: either the Jaguars would advance to the state finals, or their season would end on the court this morning. The pressure felt intense, but hopes were high!

Although fatigue was setting in for the youngest basketball fans, everyone stayed. The crowd knew the tables could turn at any time. The game had become a real nail-biter.

"Last chance for the Jaguars," the commentator announced. "Ariel has the ball; all eyes are on him. If he scores, the Rockets are through."

The cheerleaders began to cheer for Ariel. "Ariel all the way! This is your day! Make us proud and bring the crown!"

Ariel began to sweat heavily. The court seemed like a war-zone, and he was the lone ranger. The Rockets were all set to make his moves fail.

He began to dribble the ball, and the opponents began to move toward him. The pressure increased. A player tried to steal the ball, but Ariel dodged him. People in the audience were sweating.

Another player attempted to make a block, but Ariel moved around him, all while keeping expert control of the ball.

He overtook the player so that the whole crowd came to its feet.

This move awed the cheerleaders.

As Ariel moved closer to the basket, with no opponent in sight, everyone held their breath. Drops of sweat began to form on Peter's forehead. It was like him playing in place of his son.

All he could remember was his high school tenure when he was also the star of his school's basketball team.

Ariel took a deep breath and threw the ball toward the basket.

The ball was in mid-air, and no one flinched their eyes from the ball.

A player from the Rockets team came out of nowhere, jumped, and hit the ball with his left hand, making it go elsewhere. Just before it could go inside the basket.

"And it's a basket missed."

The crowd went silent. No one was cheering, not even the winning team.

Both teams began to push themselves until the referees broke the tension between them. The whole time, it was like they had turned the tables.

A final decision was made and given to the commentator.

"It was a close call. Jefferson Jaguars couldn't make it to the state finals. Congratulations, Rockets!" said the commentator.

The cheerleaders from Roosevelt High School began to chant for their team, and the crowd supporting the Roosevelt Rockets broke into clapping with wild excitement.

The winning team celebrated amid confetti.

They played their school song in appreciation, which echoed through the halls of Jefferson High School.

Ariel remained on the court in shock. He couldn't believe he was so close to scoring to move forward to the finals.

In the chortling crowd were his parents who pushed their way past everyone, down to the court where Ariel stood.

People were lifting the player who was able to stop Ariel from scoring a basket onto their shoulders. It was a big win for them.

The Jaguars were aghast. They made their way to the stands, some of them sipping water, some of them angry, while one of them began to cry.

Ariel was inside the court, eyeing the Rockets hero. It could've been him, but it looked like a beautiful dream now—far from reality.

"Ariel! Dear! Are you okay?" Jessica shook her son, bringing him back to reality. "It is okay, Ariel, it happens. We are proud of you!" His mother hugged him, but Ariel didn't say a word.

"Talk to me, my champ!" said Peter.

"We. Lost. Because. Of Me." He then broke into tears. His cheeks flushed red. He was devastated and embarrassed.

He couldn't make his parents proud; he couldn't provide a win for his team, or pride for the school, and couldn't find an answer for this setback in a game he felt so passionate about. Basketball was everything to him, and it was something that he practiced almost every day.

However, it was just the match that had ended, not the criticisms. The teammates who envied Ariel had a perfect chance to attack.

Since they still considered Ariel a hero for bringing the school team to the semifinals, they wanted to turn his fame into ashes.

"We lost the game! Just because of you. You are worthless!" said Albert.

"You did it on purpose! You wanted to have fame all for yourself. You don't deserve to be the captain of the team!" added Joel.

Ariel buried his face in his hands. For some time, he thought they were right; he didn't deserve to be in the team, and he stepped down as the team captain.

"You guys are right; I don't deserve to be the captain, I'm stepping down," said Ariel with a heavy heart.

"Come on, guys; you're exaggerating. We are a team, and we all lost this game. We can win this next year. It was because of him we even made it to the semifinals," said Daniel, trying to send them away.

Daniel was another teammate. It was he who cried on the defeat. He stood up from his seat, making his way into the basketball court. He patted Ariel on his back and said, "You did a splendid job, Ariel. Do not listen to them. There will always be people who will try to invalidate you. You are my hero!"

Albert came forward to hit Daniel for taking Ariel's side; however, the coach intervened and stopped Albert by taking hold of his hand.

"Enough, Albert. Go home! Your father and mother must wait for you and your brother!" said the coach.

He couldn't afford more drama in the basketball court, especially that they hosted the event.

Albert tried to say something but couldn't. It was like he couldn't muster the courage to say so, and he left with a sheepish expression on his face.

Once Albert and Joel had left the scene, he approached Ariel and gave him a warm hug. Tears ran down Ariel's face. The coach said, "It was a good game, Ariel. It was just too close. We had a good season!"

Ariel nodded. He didn't say a word.

The coach continued. "This is just the beginning. The next season will be much better. Trust me! Let nothing or nobody makes you lose focus. This is your passion, and because of you, we made it to the semis. A good player blooms everywhere. Nothing can overshadow talent."

"But ... look how I missed the shot! If it happened once, then it can happen twice," said Ariel. Albert and Joel had crushed his confidence.

"Look, Ariel, everything will be okay! I dreamed that one day, one of my players would be a professional player, and I see that potential in you!"

"Oh! Really?"

Someone interrupted their conversation. It was Joel. He came back to take one last verbal assault on Ariel.

"Coach! He is nothing. Our team is being trashed because of him!"

Coach interrupted before Joel could continue. "This is not the way you treat your teammate. You all owe him for bringing the team to the semifinals. It wasn't before Ariel that the team could think of reaching the semis until the last season." He didn't let anyone speak, but he continued, "He scored more baskets than all of you. Just because he lost a basket doesn't mean you all may say whatever you like. Do you have anything else to say?"

Joel's mouth opened, but then nothing came out.

"Off you go!" the coach said.

He turned to Ariel once again. He had half dried tears in his eyes, but this time, they told a different story.

He felt empowered, and he was happy someone took a stand for him.

The coach patted him on his back and said, "You will have more opportunities in the next season to show the best of yourself! It's only a bad day and not a bad life. This will only strengthen you and bring the best out of you. This is what every good player goes through, and the ones who do not give up turn out to be the best!"

While the coach talked to Ariel and comforted him, his parents saw how he was broken and that he needed support.

Peter, his father, who was a public accountant, greeted the coach and put his hand on Ariel's back. He said, "Ariel, my son, the game isn't always about winning. What you learn from it is what matters the most. You and your team did a remarkable job. Reaching the semis wasn't an easy job, but you all pulled through."

Ariel nodded but didn't say a word.

"Not that you are a bad player; no one can take the blame. The opponents were remarkable too, and we all must acknowledge that. Everything will work out fine for you next time, my superhero." With this, Peter hugged Ariel, and so did his mother.

Ariel wiped his tears and said, "I will never play basketball again! I just can't."

On this, his mother began to shake her head while Peter replied, "We will talk about this further, but for now, let's move on."

Ariel said goodbye to the coach and went to his parents' car.

He sat in the backseat and put his seatbelt on. While his parents said goodbye to the coach, some teammates who were leaving began to mock him.

"It's because of Ariel that we didn't advance to the finals," said Vincent, another teammate who practiced with Ariel in his free time.

"We could have won this game, but I just realized we have a loser on our team," added Luke, another player.

Their laughter and words were like daggers to Ariel. When they saw his parents approaching, they fled.

Peter and Jessica tried to comfort him, but it was all in vain. He didn't utter a word. On their way home, they passed by Williams High School, where a board read:

Next Friday
The Story of Challenging the Unknown
Do not miss! You can invite your whole family.

Ariel's father slowed down the car and asked one student, an ambassador for the event, about the flyers in his hands.

"Hello, young man! How're you doing?" asked Peter, trying to start a conversation with the student.

"Hello sir, I'm doing well. Thank you for asking. We are representing Williams High School. Our school will host an event called 'The Story of Uncle P: Challenging the Unknown.' The stories are not only for the youth or for the elderly. Neither are they only for men or for women. They are for everyone, the entire family. Would you like to attend?" He was eager to secure more audience for the event.

"Who is Uncle P, and what kind of story will he be reading? Is this a real-life story? Something related to Uncle P himself?" asked Peter, intrigued by the young boy.

"Uncle P is the new Spanish teacher in our school. He likes to tell fables and make people of wisdom understand the purpose behind adversities and the notion of life. He will be our chief guest for the night. If you bring your family along, none of you will regret it. You have my word."

The student wanted to be as persuasive as possible, so he added, "Also, it is free!"

Peter gave a faint smile for his effort and asked for three tickets.

The student rejoiced, and handing over the tickets said, "Do not stop attending. You all will have a great time. I assure you!"

"Yes, we will attend the event. I would love to listen to what Uncle P has for us," said Jessica.

"Thank you for your kindness!" said the student while waving a goodbye as Peter began to turn the steering wheel.

Before stepping backward, another student handed Peter more tickets for other people who might want to attend.

"Cheer up, Ariel. This story will interest!" said Peter.

"We will have a good family time, my son," added Jessica.

"I am not interested!" said Ariel.

The color in Jessica's face drained. She knew her son was devastated, and the fact that she couldn't help him made her feel terrible. Ariel's mind hadn't left the basketball court, and Jessica had to do something about it. She said, "My son! I am worried about seeing you so heavy-hearted. You have always been a good player, and you were a good player today. It is only a bad day. Even if someone or no one applauds you, it wouldn't make you less of a player. You will have your share of stardom and your popularity because of your talent. I know that will happen soon."

Ariel didn't reply. Seeing this, Jessica lost the courage to say anything else. None of it was working.

She zipped her lips until they reached home. Upon arrival, Ariel shut himself in his room. He began to leaf through his training photos, taking him back to when he was training for basketball. The album also had some pictures of Ariel with famous basketball players. Tears ran down his face; he looked at the pictures again and again until he fell asleep.

2

In the Parallel World

Sofia fastened Anthony and Leslie's seatbelts and headed to the park. Sofia left the Hamilton Street to ditch the traffic, but the street was closed by the police owing to the safety of the people attending a basketball match being held at the Jefferson High School basketball court. The people filled the court with the noise of a huge crowd watched the match between Jefferson Jaguars and Roosevelt Rockets.

"My babies, we will have to turn around. The street is closed," said Sofia.

"But why? Why are the streets closed?" asked Leslie.

"Wow! That parking lot is full of cars. There must be something exciting going on!" added Anthony.

While the cars excited Anthony, the closing of the streets disappointed Leslie. She couldn't reach the park sooner.

"Can we stay for a while? Can we? Can we?" Anthony was excited.

"No dear, we cannot. There is no parking in these blocks, and we would have to walk. With this swarm of human beings, I don't see it fitting. Also, do not take off your seatbelts while I am driving. Understood?"

"Aye, Captain. Let's go!" both the kids said in unison, and all three of them laughed while Sofia turned the car around.

After turning around, she got away from the traffic and the bustle of the Jefferson basketball court. She turned to another street with less traffic, and it took her back to Hamilton Street.

She saw a poster in front of Williams High School that read:

Next Friday
The Story of Uncle P–Challenging the Unknown
Do not miss! You can invite your whole family.

Sofia slowed down her car and asked one student about the tickets and the event.

"Hello, young man, can you give me more details regarding the event?" asked Sofia.

"How are you doing, ma'am?" replied the student.

"My name is Sofia Martinez, but you can call me Sofia," she responded.

"Sofia, the event is next Friday. We all shall gather to listen to Uncle P's fables. Uncle P is our new Spanish teacher, a motivational speaker, and a remarkable writer. He has power in his words. His stories have something to teach, not to just one age group or one gender, but to the entire family. The tickets are free of cost, and we will serve refreshments. You and the kids will have a good time. You have my word."

"Do you want three tickets to bring your children along?" the student asked in a weak voice.

"Oh! These children in the car? No, they are not my children. I would have been the happiest mom if they were," said Sofia with a heavy-heart. Yes, they weren't her children. Sofia Martinez was just their nanny, a forty-year-old Dominican woman who had been taking care of Oliver Anderson's children for over two years. She has widowed at thirty-five, and now Anthony and Leslie were her only reason to give life another chance.

She continued, "I'm their nanny, and I take them to the park every morning. We couldn't make it to the park. Thus, we're going back home. Anyway, I'd like three tickets. Let's see who can accompany me to the show."

"I am sorry for confusing them as your children. Here are your tickets. I hope to see you next Friday. I hope you have an amazing company with you. You both will not regret coming. I can assure you …" the student said. He felt embarrassed.

"Thank you for your kindness, sweetheart. See you on Friday at seven." Sofia tried to make the boy feel as if nothing had happened. She then put her car in drive, and soon, they were at Oliver Anderson Mansion, where Anthony and Leslie lived with their father.

Sofia parked the car in the garage. The passenger doors of the car opened and Anthony and Leslie came out.

"Kids! Don't run," Sofia said in a high-pitched voice.

The kids had seen their father sitting in the porch, and they raced toward him.

"I will hug Dad first!" Leslie shouted as she tried to keep up with her brother while both of them made a run.

"I will get to him first," replied Anthony as he got to the porch and hugged his father.

"But I wanted to win!" Leslie complained as she reached the porch after Anthony. She felt sad she couldn't hug her father first.

"My sweetheart, why are you sad?" said Mr. Anderson while he hugged Anthony. Then he went to Leslie and kissed her on her cheek and said, "Why does it matter my dear? Dad loves both of his children." Mr. Anderson took Leslie in his arms and hugged her tight.

"I love you, Dad."

"Dad loves Leslie more." Mr. Anderson then saw Sofia coming and said, "Welcome, Sofia. How are you doing?"

"I'm doing well. What about you?"

"When the kids are around, I am more than fine. Thanks for taking some time off to take the kids to the park."

"No, you do not have to thank me. I should thank you instead. I enjoy every moment with your kids, so I should thank you for giving me the chance to take care of your beautiful kids," replied Sofia.

"You are a sweet lady."

Sofia blushed.

"Dad, let's go inside!" Anthony pulled Mr. Anderson's sleeves.

"Yes, why not dear? You go inside, I'll follow."

Anthony and Leslie ran inside.

"Come with us, Sofia," Mr. Anderson said.

"Thank you for inviting me, but I need to go. I have a few commitments for the afternoon." She didn't want to feel uninvited in the "family hour."

"No, please don't leave so soon," Mr. Anderson impressed.

"Okay, but only for a while, just because you are insisting, Mr. Anderson," replied Sofia.

"Thanks for your consideration," replied Mr. Anderson with a faint smile.

Mr. Anderson then went inside, and Sofia followed. Just like any gentleman, he opened the door for Sofia to walk in.

"Thank you, Mr. Anderson."

"My pleasure, Ms. Martinez."

As they walked into the living room, they saw Anthony and Leslie jumping on the sofas.

"My babies please get off the sofas. Don't jump; you may end up hurting yourself," Sofia said with great concern.

"Sofia. You. Are. Staying. We. Are. So. Happy," the kids said in intervals, and they jumped on the sofas. The kids loved Sofia as much as she loved them.

Sofia smiled, and Mr. Anderson stopped the kids from jumping. He first went to Anthony, held him with his hands, and told him to get on the floor.

"But Dad—"

"Please, my child."

"Ah, okay!"

Anthony came off the sofa, though he didn't want to.

"Leslie?"

"Dad! But this is fun. Why don't you also join us?" replied Leslie. She didn't want to stop, and she kept jumping.

"Please, Leslie, Aren't your papa's best daughter?" interrupted Sofia.

"If Sofi says so ..." Leslie hopped off the sofa.

Mr. Anderson smiled and kissed his daughter.

"Children are the best creation of God," said Sofia.

Anthony, who said, "Dad, what do we do now?" Interrupted the conversation.

"Go to your room and see your new toys. Go check them out and bring them here."

Both of their faces showed a rush of excitement.

"Yay! Thanks, Dad!" Anthony made a run to his room.

Leslie followed her brother to the bedroom.

The kids soon went out of sight. Mr. Anderson offered Sofia to sit. He said, "I'm sorry. I didn't even ask you to sit. Please sit."

"Sure." Sofia sat on the sofa.

"Would you like some juice?" asked Mr. Anderson.

"Yes, please."

Mr. Anderson dialed the extension for the kitchen and asked them to bring fresh juice.

The maid came in with a pitcher of fresh orange juice and two glasses. Mr. Anderson poured juice for both of them.

They both had a good time. They laughed and shared stories. Sofia told him about her time with Leslie and Anthony and how the kids tried to make her stay just another hour.

Mr. Anderson listened to everything that she said with great intrigue. He shared stories of his secretary and his associates at his law firm. He also told Sofia about his struggles of putting the kids to bed when she is not around.

They both laughed at how Leslie and Anthony tried to avoid eating vegetables and how much they enjoyed jumping on the sofas.

"I must confess, it had been a long time that I haven't chortled like this. All thanks to you, lady. I feel so alive; this conversation made me realize the difference between living and existing." Mr. Anderson felt happy; his face brightened.

The kids were not coming back. They liked the new toys so much that they played with the toys rather than joining their Dad and Sofia in the living room.

"It is important for everyone to take some time out for themselves. Maybe two hours only can help you enjoy life because a joyful heart makes a face cheerful, but a sad heart produces a broken spirit. Laughter is the best medicine; even I laughed whole-heartedly today. I too had a good time with you. Thank you, Mr. Anderson!"

"Thank you for spending time with us, Sofia. Since the death of my wife, I have never sat downed and spend quality time with a beautiful lady," said Mr. Anderson. His wife had passed away from cancer three years ago, and the grief had been in his heart ever since.

Sofia had shared much of her responsibilities of looking after his kids, but at the end, he had to play a big role, since Sofia could not be around now and then, even though she tried her best to be.

Oliver Anderson was forty-five years old, but his loneliness made him look older. He couldn't let go of his wife's memory of taking her last breaths in his arms. He felt afraid to start over.

Since Mrs. Anderson departed from this world, he became wary of spending more and more time with his loved ones. Since the death of his beloved wife, he knew it's just in a flash of a second that the heart stops beating.

"Wow! Are you serious?" said Sofia.

"Mm-hmm!" Mr. Anderson replied.

"Excuse my boldness, Mr. Anderson, but can I ask you a personal question?"

"Go ahead."

"In the environment that you are developing, you must have met that someone special who has caught your attention. I mean, you must have met amazing women as a lawyer, as an entrepreneur, and even in your social circle. Why haven't you considered starting a new relationship?"

Mr. Anderson was silent at first.

"You may not answer this if you don't want to. That's fine!"

Sofia thought she had crossed the line, but then Mr. Anderson replied, "No, I shall answer it. I considered starting a new relationship, but I like to keep my professional life separate from my personal life. In a competitive environment, and since I am a lawyer, I have to be wary of the woman I choose since I would never know her true intentions and what she has in her heart."

Mr. Anderson continued, "People can misinterpret it as an opportunity for a promotion. I think it will be inappropriate if you run a law firm. Also, this is my preference. I don't want to come home to a wife with the same profession or a profession that stops her from prioritizing her family first."

"You've put much thought into this!"

"Yes, I have. I also don't want my wife to work for someone else when I can provide my family with the necessities of life. You know, among all, what I fear the most?"

"Oh, what?" Sofia had been carried away with whatever Mr. Anderson poured on this sensitive topic.

"I fear not finding a woman who would love my children as my wife did. I think this is my biggest fear. Among everything else, this is the biggest reason I have not considered bringing that someone special in my life and my kids' lives."

"I am sure that God has reserved someone, somewhere special for you in this world. All you need to do is standing with open arms when that opportunity comes," Sofia said.

"Thanks. I will keep this in mind. You have an optimistic outlook on life. I would like to know a little more about this optimistic woman," said Mr. Anderson with great interest.

"Sure, ask away." Sofia was curious to know what Mr. Anderson would ask.

"You are a young, elegant, and attractive woman. Don't you have that someone special in your life?"

"Um! I had that someone special in my life once. He was the love of my life, and I lost him in a car accident. A negligent drunk driver destroyed my entire world. He was everything to me; he was my world. The way he left... We didn't even have time to say goodbye. Since then, I have stayed alone." Sofia had tears in her eyes.

"Sofia," said Mr. Anderson. "We attach ourselves to someone, and then they leave us. They become a memory. A beautiful memory. But you know what the worst part is? That they only exist in our memory."

Teardrops ran down Sofia's face.

"I didn't mean to make you sad. I'm very sorry that I made you recall everything." Mr. Anderson felt bad upon making Sofia revisit her past.

"No, Mr. Anderson, he was one of the best things that happened to me. Though I miss him, but missing him is still beautiful and worthy ..."

"You are a different lady. Why don't you give yourself another chance? You deserve to lead a happy life. Why haven't you thought of forming a home and having children?"

Mr. Anderson could see the tension on Sofia's face. It seemed like the blood had drained from her face.

"Mr. Anderson, can we continue this conversation on another occasion? I need to go."

Mr. Anderson knew it wasn't time to stop her. Sofia was uncomfortable answering the questions. He nodded.

Sofia had her reasons. She could feel her heart pounding in her chest. She knew that someone would ask this question one day, and she wouldn't be ready to answer it.

Sofia got up from the sofa, and she wanted to bid him goodbye as soon as she could. Mr. Anderson was embarrassed for making her uncomfortable. The atmosphere had an air of unease.

"Sofia, excuse me ..."

"Yes, Mr. Anderson?"

"I did not intend to hurt you or make you uncomfortable, but I still did it. More than once. I am very sorry."

"Do not worry, Mr. Anderson. You just had a few questions. That's it. Everything is fine. It was a pleasure for me to spend some time with you and your children. Will you please call Anthony and Leslie so I can give them a goodbye kiss?"

"Sure. Wait, I will call them," replied Mr. Anderson.

Rather than calling them or sending someone to bring the kids, Mr. Anderson went himself. He found that fitting.

Sofia was waiting in the living room when Anthony and Leslie came running.

Mr. Anderson followed, walking at his normal pace.

"Sofia, please don't go. We want to play with you." Leslie held her hands and pleaded with innocence.

Sofia kissed her cheek and said, "My darling, I have to go. But I will come back Monday."

"Promise?" said Anthony.

"Yes, promise."

"Okay, then you may go." Leslie made a sad face.

"We love you, Sofia," said Anthony.

"Sofia loves you guys more," said Sofia as she got on her knees and hugged both the kids.

After giving love to the kids, she headed to the car.

Mr. Anderson and the kids followed. As she sat in her car, she waved goodbye to all of them. The three of them waved back.

While Oliver waved her back, he kept thinking about the questions he had asked her in the living room.

*

In a dimly lit room, sunlight shone through the glass walls. It was a beautiful office.

While two walls were made of concrete and had beautiful gray bricks on them, the other two walls were made of glass and allowed the sparkling sunlight in.

Everything was arranged on the table in perfect spacing. The table had everything in a square shape and only in black and white.

From the mug to the notepad, to the laptop and the furniture—even the interiors were a combination of the two. There was a cozy black chair with the table. A man came walking toward the chair and sat down. He was wearing colors. A blue business suit this time. He sat on the chair and began to read from his laptop screen with great interest. Keeping his eyes on the screen, he moved his fingers across the files that had been arranged on his table neatly. All the files were black; and the printed pages inside it were white, just in sync.

Ring, ring.

Before he could pick a file, the telephone extension in his office rang. Rather than picking up a file, he picked the phone.

"Jeffrey Scott speaking."

"Hello Jeffrey, this is your manager calling."

"Good morning, sir. How may I help you?" replied Jeffrey.

"I need you to come to my office as soon as possible. I have a few assignments for you to be completed in the next few days."

"Sure. I'm coming to your office right away," replied Jeffrey.

Jeffrey hung up the phone, got up from his chair, and walked out of his office, toward his manager.

Knock.

The manager saw Jeffrey through the glass door and signaled him to come inside. As Jeffrey turned the doorknob and opened the door, his manager invited him to sit.

"Come, Jeffrey; have a seat!"

"Sure, thanks." With this, Jeffrey sits on one seat facing his manager.

"How have you been, Jeffrey?"

"I'm fine. What about you?"

"I'm fine too."

"No news as yet; everything is marching in order, huh?" said the manager.

"I am glad that everything is going as per your wishes," replied Jeffrey.

"First, I would like to congratulate you for the excellent job that you are doing!" said the manager, on which Jeffrey smiled.

Jeffrey had a successful career as an English journalist. He had worked for twenty years for the Freedom Newspaper of London, England.

Now he was working for the newspaper, The Light of Washington, and had moved to Washington, DC. This was his fourth year with this newspaper and publication house. A professional and witty journalist, his interviews always left the audience talking for a long time, and his words buzzing among the masses. He had a reputable name in his field and had a legacy of keeping his bosses happy. This instance was also an example of that.

His boss continued, "All the staff gives a very good reference for you. They always have something good to say about you. Everyone keeps you in high regard, and they all tell stories about how your experiences have helped them all along."

"I'm glad," added Jeffrey.

"The company is also happy about the work that you are doing. Each interview that you conduct leaves people with a greater urge to know more regarding the issues they highlight you. Also, here you go, with a gift from the company. A gift for a job well done." Saying this, the manager offered Jeffrey an envelope.

"A gift? I hope this won't get me in trouble."

Jeffrey still hadn't taken the envelope, on which, the manager kept it in front of him and said, "Well, if you don't want it, then I can get it deposited into my account. I also don't vouch for anything bad either."

Upon this, Jeffrey responded, "One moment, sir. I didn't say that I don't want it. I just want to make sure that it won't cost me in near future that is just my concern. Thanks for the recognition and the bonus, anyway."

"We also have some assignments lined up for the next two weeks. We considered sending another journalist, but since he wasn't available for the dates, so we are relying on you for a job well done."

"Okay, so what do I have to do?"

"We require you to write two articles and an interview."

"Fair enough."

"The first one will be at a basketball match. The Jefferson High school has a match, and you have to attend the match and write to motivate parents to get involved in the sport and show support to their children who have the sporty knack."

This was before the basketball match that took place at Jefferson High School. Jeffrey would cover the same basketball match in which Ariel took part.

The manager continued, "The second is an interview with Mrs. Amanda Wilson. We all know that you have a healthy relationship with her. The interview will cover her success as a businesswoman. The purpose of this interview will be to urge young women to take up programs offered by nonprofit organizations who strive to develop more independent and successful women in the region."

"I'd love to," Jeffrey added. He appeared more interested in the second assignment.

"The third assignment is an article about an event taking place next Friday, followed by the basketball match. It will hold this event at Williams High School, called 'Challenging the Unknown' by their Spanish teacher, 'Uncle P.' He is a mentor and a motivational speaker. He tries to inculcate the spirit of optimism and positivity among masses."

"He sounds impressive," said Jeffrey.

"He is. He has words of wisdom for anyone who values it. You will also have a good time in that event. It is not only an assignment, but it has educational benefits." With this, the manager smiled.

"I will keep this in mind," said Jeffrey as he smiled back.

"I would wait to know what you have to say about Uncle P and his story," replied the manager.

"I would like to share my valuable experience with you, sir. Also, this sounds intriguing. You can count on me!"

"Thank you so much! I knew you would do it." With this, the manager got up from his seat and so did Jeffrey. Then he added, "I will have to retire for the day. I have a few commitments to attend to. Don't forget, I'm counting on you."

Both of them shook hands, and Jeffrey took his leave. Once he returned to his office, he began to organize things. Known for a job well done, he always did his homework, so he began to plan on his assignments. He updated his to-dos, dropped an email to his designer that he would require a suit for a special meeting and then; he took his phone in his hand and looked for Amanda's name. He called her, but no one answered the call, so he left a message.

"Hi Amanda. How are you? I hope you are doing well and that you have had a good trip while coming back. Please call me when you're free, and also, if you don't have any other commitment, I'd like to invite you to dinner. I'll be waiting for your call. Take care."

3

Too Difficult to Start Over

A wrinkled hand took out the cell phone from the pocket to check the time, only to find it didn't have enough charge and was thus switched off.

"Oh, oh!" She put the cell phone back into the pocket of her jacket.

Amanda then realized she was also wearing a wristwatch and didn't need a cellphone to check the time. It was a beautiful diamond-studded watch. The watch said one-fifteen.

Maybe it wasn't just to check the time, but she expected a message, or maybe a missed call—some kind of sign that people hadn't forgotten her.

Amanda came out of the airport only to find out there was no one to pick her up.

"Everyone must be busy," she assured herself.

It was a Saturday afternoon, and she saw most people coming out of the airport and hugging.

"Saturday is a weekend, and must of the people are not busy, nor are they at work." Amanda's face darkened.

"Lady, you need to book yourself a cab," she told herself. After a business trip to New York City, Amanda was home. She reached the front of her mansion in a taxi. Her maid stood at the door with a broad smile on her face. Upon seeing her maid happy on her arrival, Amanda's pale face also turned into a faint smile.

She got out of the cab and gave the cab driver some dollar bills. "You may keep the change; and please hand my suitcase to my maid."

"Sure, have a beautiful afternoon."

"You too."

Amanda went inside the house, and the maid followed with the suitcase. "How are you, ma'am?"

"I'm fine. How are you, Elise?" replied Amanda.

"I am fine, ma'am."

The conversation dropped right there as Amanda went upstairs.

Elise left the suitcase by the stairs from where another servant took it toward her room. Elise went into the kitchen to cook something for her.

Amanda observed the walls as she turned left into the hallway.

The first room was her bedroom. She opened her purse, took out the keys, and unlocked the door. She went inside, and the servant followed.

"You may leave the suitcase here and go," Amanda instructed the servant.

"Okay."

The servant left the room, closing the door behind.

Amanda sighed. She was alone. She looked around. The room only advocated silence. However, Amanda could hear a faint laugh. It was familiar. It came from the balcony. She went to see who was there. When she opened the door, she could see a misty image of her younger self.

She looked happy. No wrinkles, no dark spots. Amanda at twenty-seven years old. She glared at her in astonishment. Why was she giggling?

Then she saw another misty image. This time it was a man. She could recognize him. It was her ex-husband. He was also young. He came toward the young Amanda and kissed her on the forehead.

Amanda held his hand and said, "It will be a girl, honey."

"I couldn't be happier," replied her husband.

Seeing this, the fifty-five-year-old Amanda also smiled. She lived that moment once again, but all of it disappeared when someone knocked on the bedroom door.

The knock on her door brought Amanda back to her senses. She began to see here and there. It all had disappeared. She couldn't understand. It was only in her head. She was seeing her past; it was just a flashback. Tears began to roll down her eyes. This was when Amanda and her husband were expecting their first child.

Someone knocked again.

Tears rolled down her face. "It's all gone." Amanda sighed again.

"Ma'am? Food is ready. I wanted to call you for lunch," said Elise.

"Okay. I will come out in a while. You may go." She wiped the tears and went out of her room to the balcony. She stood at the same place where she saw twenty-seven-year-old Amanda standing. She then saw her hands. They were all wrinkled, and the skin wasn't as tight as before. She had aged.

"Amanda! It was many years ago; until when you will live like this?" she asked herself.

She herself had the answer to it and no one else. It was up to her to start a new life, but how could she? She had lived with the love of her life for twenty-five years. They had two beautiful children. Everything in her life was picture perfect until she found out her husband was interested in another woman. All those years, when she devoted herself to her husband, children, and career—what did it turn out to?

Her husband left her for a woman younger than her. He didn't even try to explain things to Amanda or complete the procedures. It was a week before Christmas when he left for California, saying he had to meet an important client. He said they would negotiate a deal over a week's time and he would be back by Christmas Eve.

That time never came. The next day, after Mr. Wilson left, Amanda received a letter from the family court. He had summoned her for her divorce proceedings. Mr. Wilson had filed for divorce from Amanda. The court citation devastated Amanda when she received this notice. She tried calling Mr. Wilson, but he didn't pick up. He didn't come back. Amanda called, but he didn't reply. Amanda's heart was shattered into pieces. She had been planning a welcome surprise for her husband, and he bid her goodbye like this.

Jennifer and Lawrence, their kids, called their father, but all that they received was a text saying, "Your mom has achieved that great height of success on her own. Thus, I want nothing from the business or the riches distributed. I'll take my fair share of the business I established. Tell her to appear in the summons. I will meet her there."

This was the only reply that Mr. Wilson had sent. Amanda didn't get the hang of the situation until she went to the court proceedings. She didn't go there to get the paperwork completed but to find out why Mr. Wilson left her. What compelled him to take this decision?

When she reached the civil courts, she realized it wasn't an abrupt decision. The reason had come along with her soon-to-be-ex-husband. She had also joined the court proceedings.

Mr. Wilson had an illicit affair with a girl half his age. They had been dating for three years now and wanted to marry each other. This was why Mr. Wilson filed for a divorce.

He appealed to the court to make the procedures quick. He said the marriage he had been in was an empty shell and he felt disconnected with his wife.

Amanda's heart sank. She couldn't believe how someone who kissed her a sweet goodbye could be a foe. Not for a single moment did Mr. Wilson make her feel like she wasn't enough for him or if she was unfulfilling. She realized how fake he had been throughout this time. How could someone's youth destroy everything that they had?

Tears rolled down her face. She tried to make eye contact with Mr. Wilson, to see if he could say all of that to her while looking at her in the eyes, but he didn't look at her. He said it all like it was a speech and left the stand.

Amanda wiped the tears off her face, got up, and said, "Your honor, I would request you pass the orders as soon as possible. I don't want to live with this man anymore." It all ended, just like that. Amanda couldn't believe it all had happened. She was a strong woman, but all strong individuals have their limits.

Knock.

"Ma'am, I have served the food at the table," said Elise and then she left.

Amanda was standing in the balcony, thinking. "At least someone cares about me," she told herself.

She went to the washroom to wash her face. After washing her face, she continued to stare at herself in the mirror.

"You need to stop living in the past, Amanda." She closed the tap and came out with a towel in her hands. She put the towel aside and went downstairs for lunch. She was the only one at a fourteen-seat dinner table. There was no one to join her. She looked all over the table. Then she closed her eyes. She could see her ex-husband and her kids sitting on the table, and Lawrence saying, "Mom, come fast, we are all waiting for you."

Clink.

Elise set a jug of water on the table.

Amanda opened her eyes. "Phew." With this, Amanda let out a deep breath and sat on the table. She then began to eat. Tears continued to roll down her face as she ate.

Elise knew how Amanda felt since Mr. Wilson left. "Ma'am, would you like to talk?"

Amanda looked at Elise, gulped her food, and said, "Sit, Elise. I have much to tell."

With this, Elise sat. "None of my kids cared to call me or pick me from the airport. I have two grandchildren, but look around. How often do you see Charlotte and Matthew playing in this house?"

"Ma'am, your kids are setting up a business of their own. Just like you, they must be busy," said Elise.

"Yes, you can engineer all the excuses that you may like."

Elise began to look down.

"And why wouldn't you? Those kids have always been dear to you. You were among the first persons who embraced them when they were born. I can understand. Look, Elise I am alone. After taking care of all the responsibilities—as a wife, as a mother, and as an entrepreneur—this is what they have left me with."

"Ma'am ..." Elise had tears in her eyes.

"You tell me. How can success and wealth mean anything to you when you have no one to share it with? For all that I have achieved in twenty-five years, this is how I am being paid back. I am alone. My companion is loneliness and nothing else. Where did I fail? Why did this happen to me?"

Elise didn't have a reply. Amanda left the food as it was and went to her room, crying. As she opened the door of her room, she could hear her phone ringing. She picked up the phone to see who was calling. It was Jennifer, her beloved daughter.

This put a faint smile on her face. She picked up the call. "Hello, my darling!"

"Mom, how are you? How was your trip? I am sorry we could come to pick you up."

"That's all right, Jenny. My trip was fine. How are you? How is Matthew?"

Jenny's call changed Amanda's mood. "Everyone is doing well. Your daughter and your grandson are missing you so much. We want to come over. Can we?"

"Sure," replied Amanda. She yearned to see her daughter and her grandson, but her reply didn't show much excitement.

She could expect how this meeting would go about, so she looked forward to both. Meeting with her kids would become a challenge because they will pressurize her into leaving her past. This argument would continue if Amanda keep disagreeing with her kids.

Her son Lawrence stopped visiting her. He put a condition in front of Amanda that if she pledges to leave off the memories of his father and start a new life, only then he would continue visiting her. Even that wasn't enough for Amanda to open a new chapter of her life. She would call Lawrence every day and tell him she missed him.

Lawrence going to the extremes stopped visiting her.

"Elise, Jenny and Matthew are coming over for supper. Please bake Matthew's favorite cookies. Don't forget to decorate them just the way he likes," Amanda said in a voice high enough to reach Elise in the kitchen.

Elise came out. She had a smile on her face. "Sure, ma'am. See, your kids haven't forgotten you. They love you a lot."

"Hmm." Amanda said nothing else.

Her daughter would visit soon. She came out of her room and sat on a sofa in the living room. Her eyes turned toward the ticking clock. She was counting minutes.

Honk, honk.

She heard a car honking in the garage. She got up in excitement. As she went toward the door, Elise had already opened it.

A chortling four-year-old came in, running. He rushed toward Amanda, who was on her knees in her pursuit to embrace him as soon as possible.

It was Matthew. Amanda embraced him, and they gave each other a tight hug. Jennifer and her husband Brian followed. They both entered the house holding hands and wearing a broad smile of their faces. Both of them were happy to see the bond between Granny and grandson.

What happened next wiped the smile off their faces.

"I am so happy to see you!" said Matthew as he pulled back. He was jumping for joy.

Amanda began to caress her grandson's head and said, "Why, Lawrence? I dropped you off to school myself. You miss Mommy?"

There was a moment of silence. Brian and Jennifer looked at each other. "Granny?" said Matthew.

"Yeah? What?" Amanda looked like she had seen a ghost. She was shocked. "What happened?" Amanda began to look here and there. She had lost the track. She had gone back to the time when she would wait for Lawrence to return from school. She missed the old days a bit too much.

Jennifer took a deep breath, gulped, and stepped forward. "Mom, this is Matthew, your grandson. I know he resembles his uncle Lawrence a lot." With this, Jennifer smiled. She tried to ease the situation.

"Oh yeah, my grandson." Amanda came back to the present. She realized what she had done. She didn't say a word. Elise was looking at Jennifer with her eyes pleading before Jenny to do something for Amanda.

"Ms. Amanda, how have you been?"

Brian came forward and hugged Amanda. "My son, I am good. What about you?"

Everybody, except Amanda, could observe a spirit of ease inside the house.

"Come, let us all go to the living room," interrupted Elise.

"Sure," said Jennifer as all of them walked toward the living room.

Amanda had arranged herself to not look like an emotional wreck, but her eyes told a different story. They were red-rimmed, and anyone could tell she had cried.

Brian held Amanda's hand as they walked to the living room. They sat on the sofa beside each other and Brian didn't let go.

"Ms. Amanda, I have missed you so much."

"I missed you too, my son!"

Brian had a special place for Amanda. He had lost his mother a few years ago, and since then, Amanda had filled that place for him.

Elise brought juices for everyone, along with Matthew's favorite cookies. Matthew started to eat the cookies and sipped the juice while sitting on one sofa on his own.

Elise, who was inclined to make Amanda's kids stay for the dinner, called her outside. "Ma'am, could you please grant me a few minutes?"

"I'm coming to the kitchen, Elise," replied Amanda.

She followed Elise to the kitchen.

"Please make baby Jenny stay over for dinner, ma'am."

"Elise, you still call her baby. Don't you worry. I will. I want her to stay as much as you want to."

Elise and Amanda smiled at each other.

During their conversation, Brian, who was concerned for Amanda, began to point this out to Jenny. "My love, didn't you notice Ms. Amanda's strange behavior? Is she okay?"

Matthew ran from one corner of the room to another. He was done with his supper and enjoying Grandma's enclosure.

"Brian, I think loneliness has gotten into her. She often misses dad, and the time spent with him."

"Jenny, this is not good for her mental and physical health," replied Brian.

"I know, my love. She loved my father too much. She loved him with all her heart. She devoted and sacrificed her entire life for my father and for us, but looks! What happened? He left her for another woman; she didn't deserve to be treated like this; it's just that the effects of her suffering have persisted ever since."

"Then we need to do something for her. She cannot live the rest of her life like that," replied Brian.

"Yes dear, I understand." With this, Jenny got up and made her way to the kitchen.

She saw Elise and Amanda chopping veggies. "What are the beautiful ladies up to?"

She entered the kitchen with a big smile on her face and hugged Amanda from behind.

Elise smiled.

"Nothing much, my darling. You and Brian are staying for dinner, so I'm helping Elise in getting the arrangements done," said Amanda, and she planted a kiss on Jenny's cheek.

"Oh, Mom. We would love to stay. Matthew always wants to sleep over at your place."

Elise slid from between them without having them noticed.

This was a perfect time for the mother and daughter to have a heart-to-heart conversation, and Elise understood this well.

"Mom, why have you brought yourself to the edge?"

Jenny asked this question out of nowhere. "What? What made you say so?" Amanda didn't understand where this conversation was going.

"Mother, now, you need to take a step forward and break the bonds of the past. You need to change your life. You have been the best mother, the best wife, and you are an incredible human being. You don't deserve to stay alone. Dad didn't give two minutes' thought to divorce you and look at you. You have been carrying his memories ever since. You need to let him go, once and for all."

With this, Amanda had tears in her eyes.

"Tell me, daughter, what went wrong? What did I do to deserve this?"

Jenny wiped her tears and said, "My dearest mother, you did nothing wrong. You are a talented and hardworking woman who took her family to new heights. This includes my father. You always prioritized your family above everything; the thing that went wrong was that you loved the wrong man. You loved someone who didn't care about feelings."

Amanda sniffed. She was crying. She kept the vegetable bowl on the side and sat on a chair by the kitchen counter.

"Mom, please don't shed tears for someone who wouldn't heed your emotions. Tell me one thing ..." Jennifer had something in her mind.

"Yes, go ahead." Amanda was intrigued by her daughter question.

"Wasn't your time with Dad a happy time? Didn't you live the best days of your life with someone who didn't value emotions much?"

"Yes, it was."

"Then tell me, wouldn't the rest of your life be harmonious and wonderful if you continue it with someone who would value you?" asked Jennifer.

"You don't understand ..."

"I do understand, Mom! Just tell me: yes or no?"

"Yes."

"Then what is stopping you?" Jennifer held her hands, and Amanda knew there was no escaping from this talk now.

"I am afraid to start a new relationship after this experience.

How can I trust someone with my heart? It is easy for people to crumble it to pieces. At this age, I have no desire to go through this again. It is better to stay alone. I may have been destined to stay like this, alone ..."

This made Jenny break into tears. "No, Mom! Please do not think like that. When I see you hustling through life and grinding on your own, a part of me dies daily."

"Every day when I go to bed with the thought I couldn't do anything for you; I feel that I have failed as a daughter. I cannot live beneath this burden, not anymore."

"No, please don't say that, dear! You have been the best daughter. I couldn't ask God for more." Amanda rested her head on Jenny's shoulder.

"There are good men out there. You just need to give a chance to the right one. Even men have gone through experiences like you and they too long for a companion."

"Please stop right here. I don't want to talk about it anymore." Amanda wiped her tears. "Matthew must wait for Gran! You've made me stay put for so long. Let's go back." Amanda tried to change the topic.

"Let us get this good mom a good husband." Jenny laughed. The mother and daughter duo left the kitchen to join Matthew and Brian in the living room.

After they left, Elise entered the kitchen with a smile on her face. She began to make the rest of the preparations for the night.

Amanda stopped Jennifer and said, "I am glad that my daughter thinks about me so much, but my child, I will be fine."

"I will make sure you are," Jennifer impressed as both of them walked in.

"Gran-gran! I have been waiting for you for a long time. Come, play with me."

Matthew came rushing and took Amanda with him. Amanda and Matthew sat on one sofa, and Jenny sat on the other by Brian.

"Is everything fine?" Brian whispered to Jennifer.

"It will be." Jennifer smiled.

"So, Ms. Amanda, tell us: how was it in New York? We want to know."

Amanda began to give details about her visit, which was educational. Brian was always impressed by what she said, and he kept asking more questions.

Because Amanda was so intelligent, she had the answer to everything. She would take-out points that no one else would think about.

"Now I know why your daughter is so intelligent," said Brian as he looked at Jennifer. Jennifer smiled.

"Yes, my daughter is unique in her ways."

"Not only that, she is a good wife and a good mother. This is also because of you. You are her role model, and I am happy I married her."

The three of them talked as Matthew played around the house. Matthew was an innocent yet notorious child. He was witty and never stopped asking questions, just like his father. He made Elise run around him the entire time Amanda was away. He played in the garden and in the kitchen. Elise, too, had a good time with Matthew. She loved Charlotte and Matthew. This was because of the bond she had with Jennifer and Lawrence.

Everyone talked over dinner and shared stories. Elise had joined them.

Matthew wished to be fed by Elise, and she was more than happy to do so. "He is like baby Lawrence," said Elise.

Soon after the dinner was wrapped, they all had coffee. "Mom, I think it is about time. Matthew has to go to sleep early, and you too should rest. We have taken much of your time. I will call you tomorrow," said Jennifer.

"I don't know if I need rest anymore. Meeting you guys and my baby Matthew has been a therapy on its own. In fact, I will head to the park for some fresh air.

"Are you sure? Wouldn't you want to take a rest?" asked Brian.

"No, I don't so," replied Amanda.

"Then come with us. We will drive you to the park on our way home."

"I would prefer to walk my dear child."

"Okay, as you like," replied Brian.

With this, Brian and Jennifer bid goodbye to Amanda. Matthew wanted to stay, so he began to request, "Can't I stay more? Everyone deserves a night off."

Everyone laughed at his innocent request.

"No, my dear child, for you going to sleep early is important. However, you can visit us again whenever you like," said Elise.

"Mom, I like Elise a lot," said Matthew, holding Jennifer's hand. He planted a kiss on Elise's cheek, and Elise looked happy.

"It was an honor to spend time with you. You aren't only a figure in my life, but you are also my mentor and my teacher."

With this, Brian hugged Amanda. He took Matthew into his arms and began to head out. Jennifer followed them after giving a brief hug to Amanda and Elise.

"Love you, Mom," said Jennifer as she left.

"I love you more," replied Amanda. Jennifer and Brian had made her day.

Amanda watched her children leave. She then closed the door behind them and passed a faint smile at Elise.

"I told you, ma'am, your kids love you more than anything in this world."

"Maybe you are right, Elise," smiled Amanda. She was beaming.

Her face was fresh. It didn't look like she had been back from a trip and hadn't slept.

"Can you please get my coat and my cellphone? I am planning to head to the park."

"Sure, ma'am."

Elise went upstairs to Amanda's bedroom as she waited in the lounge.

Elise was back within a few minutes.

Amanda put on the coat and unlocked her cellphone to check if she had any important messages or calls to attend to. Upon skimming through the messages and the calls, she finds Jeffrey's missed calls and then a message that said: "Hi Amanda. How are you? I hope you are doing well and that you have had a good trip while coming back. I hope this message finds you well. Please call me when you're free. And also, if you don't have any other commitment, I'd like to invite you to dinner. I'll be waiting for your call. Take care."

Amanda called Jeffrey back. Jeffrey was in the middle of something important and ignored the call. Since the phone kept ringing, he got up from his seat and went to the sofa where he had left the phone. Upon seeing Amanda's name, he picked up the phone and said, "Oh, I am sorry. I didn't know it was you."

"That is no problem, Jeff," replied Amanda.

"How ... how are you, Amanda?"

Jeffrey stumbled as he talked. However, it wasn't something unknown to both of them.

Amanda laughed and replied, "I am good Jeffrey. How about you?"

"Umm! I am doing fine. How was your trip?"

"Jeff, Jeff ... how about you ask these questions this Wednesday at dinner?" said Amanda.

"Oh, so will we be meeting?" replied Jeffrey. For a journalist, he was a bit too nervous about meeting this woman.

"Yes, I think so," replied Amanda.

"Okay see you on Wednesday," said Jeffrey.

"Yes, take care."

"You too."

With this, Amanda hung up the call. She then kept the phone in her pocket and went out of the house toward the community park.

***

It all started on the same day. Nine o'clock in the morning. The sky was clear, and a cool breeze blew through town. It was the perfect time for couples to head out on a romantic date. Roadside breakfast joints were occupied with couples on each table. Everyone has a way to celebrate the romantic weather, and all these ways differ from one another.

A middle-aged man, Oliver Anderson, opened the metal gate of the graveyard and went inside. He was carrying a bouquet of tulips and sunflowers in his hand. He headed to a grave on the left. The tomb stone read: "Elizabeth Anderson 1978–Forever 'Those who love us, never leave us.'"

Oliver came to celebrate this romantic day with his deceased wife. He deposited the beautiful bouquet next to the grave and sat on its side. He traced the letters on the tombstone and began to weep. He closed his eyes, took a deep breath, and then opened them.

He then said, "My love, Elizabeth! How have you been? Did you miss me?"

It looked like his wife was sitting in front of him, listening to him.

"I miss you. It has been a long time. I still remember lowering you into your ultimate resting place. I wish I could lay by your side, but life. Why did it bring me here?"

He cried so hard that he began to gasp for breath. He wiped his tears, which continued to fall, and poured out his heart.

"I can never let go the feelings that I would get when you would hug me ... how we would prepare breakfasts on Sunday morning ... our beach dates ... I cannot let go of it. You would always stand by my side no matter what; your words for consolation worked like magic. Since you've left, I haven't found solace at any moment until now. Every day, I would look forward to kissing you once I come back home. I miss those family dinners in which you would be present. Ah, you have left behind everything my love, and taken my will to live with you."

He then paused for a second.

His phone was vibrating. He took it out; it was his mother. He cut the call and replied with a text message: "Mom, I am with my love. I will get back to you ASAP. Love, Oliver."

Then he put the phone in his pocket and said, "Sorry, dear. It was Mom. She also misses you a lot. You were like a daughter to her. You were such a loving daughter to her, a responsible mother to our kids, and the best wife for me. I don't know; how did you juggle all those responsibilities? That is a lot of work for just one person. Since you've left, Mom takes care of Anthony and Leslie. That is a lot of work, and when you were around, you never made me feel like it was such a big responsibility. Mom is old. She couldn't take care of her grandchildren on her own. Thus, I hired a nanny. She is an honorable lady. However, I wish that the time to hire her would never come. I wish you never had to leave us. I wish you were there to see your children growing up. Anthony and Leslie are growing up fast. It's like we are just running out of time with each other, and this scares me a lot. Just like I ran out of time with you, I fear that this could happen with my mom and our kids. Your leaving has taken away the charm from my life."

During this soulful conversation between Oliver and Elizabeth's souls, two birds came flying out of nowhere and sat on Elizabeth's tombstone.

The birds sat close, with their heads inclined toward each other and started to sing. Oliver began to look at them and said, "Look, Elizabeth, everyone is celebrating their love in this romantic weather. I wish we both could." He began to look toward the sky and said, "God, why did you? Why did it have to be her? Why was my family chosen for this?"

Then he began to look on the grave once again and said, "Why people with a good heart don't stay around for long?" He didn't get an answer. He let out a sigh and said, "I am only yearning for responses to my questions and, you haven't responded to any. I wish you would at least come and visit me in a dream once, but you know what, I have never had a good sleep since you've left. How am I going to see you in a dream like that? It is like the sky is bearing witness to our lives. More like a silent spectator, it oversees how to respond to different situations and tragedies that happen in our lives."

During this time, Oliver talked to himself. He enjoyed visiting Elizabeth revisiting his past, all the good memories with her.

He had settled his law firm after he and Elizabeth started dating. Elizabeth had been a silent supporter during all this time.

She took her last breaths in Oliver's arms. Oliver was the one who lowered her in her grave. As Oliver closed his eyes for a few seconds, he saw a glimpse of Elizabeth taking her last breath.

She said, "You are the love of my life. I only prayed one thing: 'that I die in your arms.' I would be the happiest woman. I know God will grant me my last wish."

With this, Elizabeth closed her eyes and never opened them again. She had a faint smile on her pale face. Oliver, who expected her to open her eyes again, was shocked. He held Elizabeth in his arms as his mother cried. As soon as he heard his mother's faint cry, he opened his eyes in shock. His face was pale, as if the blood had drained. The first thing that he saw was Elizabeth's tombstone. He wasn't going through the pain of losing her again. This is losing someone you love: they don't leave altogether.

This happens in turns. First, the soul departs the world, and then it is the body, which is laid down in the grave, and then the memories.

One after another, neither are they forgotten, nor do they fade. They remain there, as it is, engraved in the deepest of our hearts.

If any of us think they would never come back, or we would never miss the deceased, we are wrong. The memories linger, and they shall linger forever. Destiny throws something related to them in our faces now and then—their birthday, anniversaries, first date, something specific to them and much more—and they all come around in disguise.

At this moment, as Oliver looked around him, he could feel nothing other than loneliness. He felt like he had lost everything. His heart felt heavy. He gulped and closed his eyes again. He told himself, "Oliver, Anthony and Leslie are waiting. You need to go home. Why have you forgotten that you have a mother and two beautiful kids? Those kids are a living advocate of the love that you and Elizabeth had. Get ahold of yourself!"

He then opened his eyes and took a tissue out of his pocket. He wiped his tears and kept it back. He then put two fingers on his lips, kissed his fingers and took it to where Elizabeth's name was written on the tombstone.

Then he said, "Baby, I am leaving now. I will come back soon. Anthony and Leslie also miss you a lot. I will bring them to meet you. It's Saturday and they must wait for me at home. I love you, a lot, forever …"

He let out a sigh, and he waved goodbye to the grave. He sat in his car and took out a water bottle and a handkerchief. He then wiped his entire face and put the handkerchief on the dashboard.

He didn't want his kids to notice that he had cried. He put on his seatbelt and turned on the car.

It was a beautiful day, a beautiful day to celebrate and a beautiful day for new beginnings.

Oliver, who was successful in his life, now feared to lose his loved ones. As a human being, it is natural to fear such a loss, but should this stop us from moving ahead in our lives?

Should it stop us from loving someone else just because we have lost a loved one in the past? Shouldn't we with open arms? Is it right to give in?

We will later find out.

4

In the Big City, All by Yourself

The weather felt cool outside. It was five in the morning, everything was quiet. However, the clock radio provided weather updates.

"The weather forecasts project that today will be a beautiful day in Orlando, Florida. The sky will be clear with temperatures between seventy and eighty degrees Fahrenheit. It will be a beautiful day to do different activities."

A young girl, age twenty-five, heard the clock radio with her eyes closed. She doesn't want to leave the bed, but work calls. She slid her legs out of bed and sat upright. She stretched once as she yawned. With a frown on her face, she extended her hand to the side table and turned off the alarm clock. She stayed on her bed and began to play with her hair. She let out a sigh. She got up from the bed and walked toward the balcony. As she opened the door, the cool breeze hit her. She stood by the railing of the balcony as began to look down.

The balcony was situated on the second floor of an apartment building and gave an incredible city view. She wasn't admiring the early morning breeze and the beautiful skyline, but she was in deep thought.

As she looked down into nothingness, she said, "Again, Ariana Tilson? The same routine? Until when?"

She was discontent with her routine. As a beautiful young woman, Ariana was not happy.

Although she had three years of experience as a nurse at Jesse Hospital and did well in her job, something kept her unsatisfied.

"Oh Ariana, you will be late for the hospital." As she blurted this out, she came back to the present. She made a run toward the bathroom.

"Jog it up Ariana. It's Saturday, the last day before you vacate." She washed her face and got into the shower.

As she came out of the shower, she didn't look fresh. This is normal among people who are made to do things they are not happy about. She began to get ready for work.

She had her uniform on, keys and bag in place, and set to leave the apartment. She had no one to say goodbye to in that beautiful apartment; she lived alone. As she rushed toward the door, the breakfast table was empty. She ate nothing; maybe she didn't feel like it. Before she closed the door, she came back and went into the kitchen. She opened the fridge, took out an apple, and put it in her bag. She left the apartment, making sure the doors were serrated.

As health-conscious as Ariana was, she took the staircase rather than the elevator. According to her, it additionally takes longer. Ariana's spirits felt lifted. It was the weekend. She got in her car and drove to the hospital. With her bag kept in the front passenger seat, she turned on the radio and her favorite song was on. Her car left the parking lot in slow motion.

On her way to the hospital, she had a faint smile on her face. The song must have changed her mood. She would soon arrive at Jesse Hospital.

"Again!" says Ariana as she reaches the hospital. She parked the car in the nearest unoccupied space. She took out her bag and rushed into the hospital. She took the elevator because she needed to report on the seventh floor, and if she would take the staircase, she would probably be exhausted. As the elevator reached the seventh floor, she stepped out and walked to the reception desk.

She signed in, grabbed her files, and reported to the doctor.

"Good morning, Doctor!" said Ariana with a pleasant smile on her face.

"Good morning, Ariana! How are you doing? Today is a good day for you; congratulations!" Dr. Judith greeted her with a big smile.

"What? Why?" Ariana was surprised.

"The department has chosen you for the employee of the month, my child!" said Dr. Judith as she held Ariana's hand. She was an old lady who liked hard workers. From the first day, she had admired Ariana for her dedication. It was Dr. Judith's personal request that made the department assigned Ariana to report to her for a month. This time, it was not just a month recognition, but the department had also granted Ariana a promotion.

Dr. Judith put forward an envelope for Ariana. When she opened it, she was awestruck. It was her promotion letter. Ariana had tears in her eyes. Dr. Judith hugged her and said, "You deserve it, my child!" Ariana was happy, but she thought just one thing: only if her family was around to celebrate this.

However, Ariana did not know what this called for. Though she had been granted a promotion, her supervisor was still not satisfied with her. He was rather jealous. As a young girl excelling in her career, he couldn't take it. He would often call her to fill in for employees, never let her take vacations, and even made her go through a hectic work schedule so she'd quit. He even went further last month when he complained to her superiors she had been neglecting work. He made up stories about how she was late to attend a few patients for a week. He even wished something tragic could affect her attendance. These false accusations and questions on her work had left Ariana disheartened. She didn't feel enjoy working at that hospital anymore.

Ariana kept herself immersed in work throughout the day. As the day ended, she left. As she entered the parking lot, she saw a group of coworkers talking.

As they saw Ariana approaching, one of them said, "Hey Ariana! Come! Join us."

"Sorry, I have to leave. I need to prepare my bags for my trip. I wish you all a good time. See you all soon," replied Ariana.

She didn't stay to talk to any of them; however, another nurse, Margaret, followed her. She said, "Come, Ariana, I will walk with you to your car." Ariana nodded.

She said goodbye to the others and accompanied Ariana to her car.

"Thanks for coming with me. I do not feel comfortable going out with some coworkers because they are two-faced. Some of them are accomplices with the supervisor. Therefore, I keep a distance from most of them."

"I understand and I agree, Ariana," replied Margaret. "How about we have dinner together tonight?"

"Okay, but with one condition ..." replied Ariana as both of them walked.

"I hope it's good. What's your condition? I will try my best to live up to it!" replied Margaret.

"That we order Chinese food from my apartment and have dinner at my place. What do you think?" impressed Ariana.

"It's not a bad idea. It's good. In fact, I like it!" said Margaret with a faint smile on her face.

Margaret used public transportation, so she accompanied Ariana in the car and together they drove toward Ariana's apartment. Along the way, they began a conversation where Ariana confided in Margaret about how she has felt at work for the past six months of how alienated she has felt lately.

It is difficult to confide our feelings even in people we trust. Ariana, thinking she cannot take it anymore, pours her heart out to Margaret.

"I do not feel comfortable at work, not anymore. I feel that everything I do is crumbled to the ground with false accusations and negative comments from our supervisor and his accomplices," Ariana said as her eyes filled with tears.

Margaret held her hand and said, "I can tell. It is no longer the same hospital as when we started working there. It seems like its favoritism for some. A few people get things the easy way, without even working for it."

To this, Ariana replied, "I want to apply for other vacancies, but I'm afraid. I do not know what it would be like to start in another department or a new job. I am afraid of change. I don't know how my new job will be. This was my first job since college. After three years of working and doing the same thing, I feel comfortable in this place. Even when the supervisor or colleagues are not good and do not want to recognize my good work in the department, I feel like I'd be lost if I left this place now."

"Do not be afraid, Ariana. You only fear the change. It is just a change of place for anyone who knows the dos and don'ts of his job. I'll confess something that I have told no one. On Monday, I have an interview for a new job. Please keep it a secret. I have trusted you with this," said Margaret.

Ariana was excited to learn Margaret, her best friend from day one at Jesse Hospital, will change jobs.

She knew how exhausting it had become in the hospital and that everyone needed a change. Margaret held Ariana's hand and said, "Cheers! To new beginnings!"

Then Ariana said, "I'm happy for you and I wish you the best in the interview; you will ace it."

After twenty minutes of driving, they arrive at Ariana's apartment.

Ariana handed the keys to Margaret who took the stairs while Ariana parked the car in the basement. As Ariana got comfortable in her apartment, Margaret took out her cell phone to order Chinese food for two. Though Margaret is not fond of Chinese food, she gave in to Ariana's wishes.

"I'll be out of state for seven days. Outside of work commitments. I feel free and I feel thrilled. My sister moved back to Maryland about three years ago. This was a few months after I started working at the hospital."

"So you will visit your sister?" asked Margaret.

"Yes!" said Ariana wholeheartedly. "We both miss each other. She wants me to look for job opportunities in hospitals or clinics in Maryland so we can stay together again. As a big sister, she is very protective and always looks out for me. She wants me to stay as close as possible."

"That is the responsibility of the elder siblings. They want to take care of the younger siblings in any way possible!" replied Margaret. "Dear, this one thing came in my mind."

"Aha, go ahead."

"I was thinking, why don't you listen to your sister? Maybe find a job in Maryland. A person is always happy when he is close to his family."

"I do not know. This may sound good, but I have to think about it."

"Think about it? Why? You just have to do some searches about vacancies when you get there. You've got nothing to lose. Promise me you will do it. Please!"

"These are my vacations and I do not want to be in the street looking for work. It is no job hunt! I want to enjoy it to the fullest."

"Either I'm on vacation, or I'm looking for a job. How can there be something in between, Margaret?"

"Girl! You can do both. I know you will have a good time with your family, and it will become much better when you know you would not leave them again," said Margaret.

"Okay, I will do it because of your insistence and because I know you want the best for me. However, I am still not so sure about it! I appreciate your friendship. After my sister moved out, I didn't have anyone close to me whom I could trust and share what I felt; you are like a second sister to me sent from heaven. Thank you very much."

"You can always count on me. You are like the younger sister I never had. I value your friendship. You are always there to listen to me and advise me whenever I need it," replied Margaret.

Ariana and Margaret gave each other a hug as gratitude, love, and consolation.

That night, after Margaret left, Ariana packed all of her stuff.

She kept all her essentials and began to think about the possibility of moving back to Maryland.

Margaret had left with a heavy heart, but she didn't want Ariana to know.

Ariana's flight was on Sunday morning. She loaded all her stuff in the cab and off she went.

***

It is difficult for people to work and study. Double jobs aren't easy. A young Afghan mother, Emma Dil, has two jobs.

She aspired to be a social worker and has been working in it for a long time. Apart from this, she is a student and mother of two, ages fifteen and nine. In the last year of college, it has become difficult for her to make ends meet.

As she talked to another student, she said, "Sometimes I've felt like giving up. I have skipped important college classes just to take care of my kids and make ends meet. I have worked extra, I have pushed myself to the edges, and we didn't have enough food at home. How did I manage? I would skip meals myself and provide for my children."

"You are a strong woman, Emma, and I know, one day, you will have all that you are working for; just don't lose hope!" replied the student.

They walked inside the university building. Emma had a presentation that day. She had to present her homework and talk about how she can help victims going through abuse or have been a victim of abuse in the past. She started with her personal life and the atrocities she faced before leaving Afghanistan. She then entered the section where she told about how she managed in the United States. She was an open book when she talked about her escape from Afghanistan, how she aspired to be free, and how she worked for it.

She continued, "Women in my country are ill-treated. Every household has a story of domestic violence, and hospitals are full of female patients who have suffered domestic violence at the hands of their spouses. The cases that make it to the court of law are thrown down the pipes when the plaintiff is dead at the hand of their husbands. To your surprise, let me tell you, the court would support all such criminals."

"Wow!" the class said unanimously.

She continued, "If I am still standing and able to quote all of this in front of you, let me tell you, this is because there is someone up there who has been watching over me and my innocent kids. Sometimes I preferred to die rather than live. I felt like giving up; I didn't see a single thing benefiting me. I wanted to end this life. What is the purpose of living with so much suffering? It made little sense to me. After seven years of emotional and physical abuse by my husband, I escaped with my two children. Why should I destroy this beautiful life at his disposal? We all may live a free and satisfying life. I waited for my husband to go to work, and with the help of some neighbors, I escaped. I fled from my home. I didn't know the job advised my husband to not go for work."

"When I thought I had walked away free, I realized that more was yet to come. When I was already near the British embassy, my husband came after me. He ran after me to stop me and shouted for people to stop me. He had crossed all his limits. I knew if he could catch me, I would end up dead. A security guard from the embassy saw me running with my children and soon, a crowd of men came running after me. I heard a sound like an alarm and all the staff mobilized to protect the entrance of the embassy. Since many did not know what was happening, they only saw a group of men running toward the entrance."

"When I arrived at the entrance, a group of armed men went out of the gate to protect the entrance from outside. When they opened the gate, it gave me time to enter with my two children. Then my husband started screaming to get me out of the embassy because I had taken his children. Some security guards took us to a room where they gave us food. They held a full-fledged investigation. They questioned me for two hours to find out why I had fled and entered the embassy."

"After five days in the embassy under political asylum, they gave us a refugee visa to enter London. We lived there for six months, and then, I had learned that my husband had relatives and friends looking for me to kill me and return my children to their father. On two other occasions, friends of my husband who threatened to kill persecuted me. Twice, I was close to death. When I could not do any more, I sought asylum in the American Embassy and after a month full of panic and terror, they gave us the visa to come to this beautiful nation that has welcomed us with a lot of love."

The class was silent, taken aback. As Emma Dil said all of this, her eyes, too, were filled with tears.

She wiped away her tears and said, "Knowing no one here, we arrived at the Baltimore airport. We did not know where to go. However, that deity in the sky oversees each one of us. Two Christian missionaries we met at the London airport noticed we were lost and offered their help. Even though my English was not good, we could have an effective conversation. They told us they were also going to the United States and they would board the same plane."

"These Christian missionaries, without knowing us, never left us in all the trajectory of the trip, including the immigration service. When we arrived at our destination, I was afraid to ask them questions because, for me, I saw them as suspects since they offered their help for nothing. When we said goodbye, the missionary's wife asked me if someone was coming to look for us. I told her I did not know anyone, and that it was our first time in the United States. I tried to explain how my two sons and I had left Afghanistan. They had a conversation with each other and then they approached us and offered us lodging. My two children and I started crying because the only thing we had was thirty dollars in cash to eat something and not knowing where we would sleep."

The class was silent.

No one could know what her smiling face hid. Can we know of

anyone's suffering by looking at them? No, we cannot. Therefore, we all need to look out for anyone who may need help.

We all have become so immersed in our daily affairs that we do not realize everyone around us is fighting a battle within their own selves.

While Emma made her presentation at the university, her eyes filled with tears and they began to flow down her face. These tears were not of sadness, but of gratitude and happiness because strangers, whom she had never seen before, will open the doors of their house and hold them up for two years until she could learn English better to become independent. She had been given a helping hand.

She continued making her presentation, saying, "This couple of Christian missionaries seemed like two angels sent from heaven, to help us and provide us with all our needs for two years. They did it free. Trust me! This has led me to think there has to be a supreme being who knows the needs of his children and sends them help at the most critical moment to help them. These two Christian missionaries changed our lives forever in those two years. My children and I have our own place to live now. Only because of them were we able to stand on our feet. I confess it has not been easy, but we are still on our feet. We have to fight a different battle every day, but I can say I am making it through. We are grateful to that couple for all they did for us. That is something that my children and I can never repay."

As most people had a mix of emotions and couldn't utter a word, a classmate stood up and said, "Wow! You have left me without words!"

Then another said, "And I thought my life was a disaster when I emigrated from El Salvador, leaving behind my family, my belongings, and my place of comfort. After I listened to you, I do not regret having made that decision because my biggest fear was not knowing what I would find during my journey and all these years. I salute you, ma'am!"

Then another stood up and said, "I do not regret having done it because now I can see the world from a different outlook."

The professor intervened. Her eyes were filled with tears and she said, "Let's not depart from the presentation. Emma Dil, if you're done with your presentation, let's opens a question section. I want to be the first to ask you a question. What has been your greatest fear all these years?"

To this, Emma replied, "My biggest fear is the fear of failure, but more than failure is the fear of that feeling that failure produces. I think we are all afraid of that part, afraid of the sensation that failure produces. We are afraid of falling apart and we are afraid of the unknown. For me, as a woman, it has been difficult but possible to reach a country that I did not know the culture and the language to break through the crowd to take my place in this society. It has been difficult, coming from a culture where women are not valued and may not have a voice or vote; it has not been easy to have this change of mind to change my life."

Another student had a question and raised his hand. He asked, "Do you believe in purpose? Why are you here in this nation?"

Emma replied, "You know, that's a question I asked myself when I was in my country. On that occasion, I wondered why I was born, in a poor family, and then my parents married me against my will. Why did all of that happen to me? Then, having two children only saw how their father mistreated their mother. Why? After having emerged from the oppression I had endured for seven years, I can now see the world."

"It is a whole new world for me, whose secrets unfold one after another. So, answering your question, yes, I think there is a purpose for every human being born in this world, and, many haves been entertained along the way, others diverted from the original purpose for which they were created, and end up being the bad guys in real life. I believe that everything finished well, according to the book of wisdom."

Another student raised his hand and asked, "What happened to the Christian missionaries?"

"They devoted their entire lives to the missions and moved to Kenya, Africa," replied Emma.

Having answered this question, the teacher came up with another question from her. She asked, "Have you ever considered returning to your country?"

To this, Emma replied, "Yes, I've been thinking about this more often, but I'm afraid to do it. There is so much to be done before I go back. And I would also like to return to help and educate women in the country. Many die of hunger because they do not know how to do anything to survive in a culture where they are not given a chance to earn their daily bread."

Her classmates agreed to help her as much as they could so she could return to her country to help women and abandoned children going through difficult situations.

With this, the bell rang, and the class ended. All the students made their way out of the class, but this time, one thing was different: everyone saw Emma with much more respect. Their eyes were full of empathy.

Emma Dil made her way out of the university. She sat in the cab she had booked and went home. As she reached home, she made herself a cup of coffee, sat back, and began to relax. In no time, she fell asleep. The coffee cup sat on the table, unattended. Emma didn't even have a sip. She was tired, and this sleep was much needed.

Emma was woken up by her ringing cellphone. Someone must have called. She put the phone on silent and went back to sleep.

The cell phone rang for the second time, on which she got up in surprise. What time was it? Had she been sleeping all along? She saw her phone. It was Sunday and two-thirty in the morning. Who is calling? In shock, Emma picked up the phone.

"Hello?" answered Emma.

"Peace be upon you and your children!"

The person over the phone spoke in Afghani.

"It's me, Nemat, your big brother, who is calling you from a land far away from yours ..." said the voice.

Emma's heart shuddered; she thought something had happened to someone in her family. Why else would her brother call her?

"How are you? Why are you calling? Is everybody fine, Nemat?"

"Yes! Everything is good," answered Nemat.

"I got nervous when I heard your voice. It is two-thirty in the morning in the United States," Emma replied. "Why are you calling? Please get to the point. You wouldn't call me at this time of the day without reason," pressed Emma.

"It is about Karim, the father of your children. He was heading to his work this morning, and the terrorists attacked. Fifteen people died, while others were injured. Among the deceased is he."

Emma said nothing.

Nemat continued. "After you left, his life was not the same. Not even a bit! I talked to him two months ago, and he told me how sorry he was for whatever he put you through!"

Emma was still silent. She didn't know what to say. It was like some kind of silence had taken over her.

"He would have liked the last chance from you, but he knew it was too much to ask for. Emma?" said Nemat.

"Go ahead! I am listening," replied Emma.

"He wanted to see his children again—to embrace them and to apologize for not being a good father. I have to leave you; I'm running out of time. I have ten seconds to finish this call. We love you and we hope to see you some day, bye. Be a strong sister. Your kids are only left with one parent now."

It was like all emotions had left her. She couldn't come to terms with the fact that Karim was no more. Emma did not know how to react. After thirty seconds, she burst into tears.

She had been widowed at the hands of the terrorists. Years back, her husband wanted to kill her, and now he had been killed. He was no more.

Though her husband had put her through so much, she never thought something like this would happen to Karim. What kind of unexplainable feeling is love?

How can we let people walk away free for what they have done to us, just because we love them? So strange, yet so precious. It doesn't happen every other day, but once a bird falls prey to it, the bird cannot do anything other than surrendering to it.

Emma lived in a one-bedroom apartment with two beds: one for her and one for her children. She put her hands on her mouth to muffle her cries. But they were loud, loud enough to wake her kids up.

Ibrahim, his eldest son, got up and asked her; "Why are you crying, Mother?"

She called him with the signal of her hand and gave him a hug.

At that moment, her youngest son Ali also woke up and listened to her mother telling his brother Ibrahim what had just happened with their father.

Both sons hugged their mother and began to cry with her.

Emma knew she couldn't go to see her husband, even for the last time.

She told her kids that they would keep their father alive in their hearts, always. They just had one picture of Karim and nothing else. Karim was soon to be buried as per religious rituals. What would be left of him in this world? Nothing he will become a memory, not a beautiful memory for them, but a memory.

"Mom, we won't ever see Dad again," said Ibrahim with his red-rimmed eyes.

"No," replied Emma.

"Can we go to Afghanistan to see him one last time?" asked Ali.

"I wish we could, my dear. I wish," replied Emma. Tears rolled down her face. "I, too, wanted to see him. I wanted to tell him that if it was not for his abuse, I would have never left him." She hugged the kids again as they wept.

They saw the breaking dawn. The family spent the rest of the day crying at home. They didn't leave the house; it was a silent day through and through. Emma had been widowed. Years back, she wanted to get rid of this man, and now, this man was no more.

They stayed awake all Sunday morning. They talked about the time when they were back in Afghanistan, with Karim.

"Dad never remembered my birthday," said Ibrahim.

"But he loved you more than me because you were his firstborn," replied Ali.

"He loved both of you. He was happy to be the father of two sons; he often told me you both are the apple of his eye." She kissed both the boys on their foreheads and held their hands tight.

The normal day started. The kids went to school, and Emma went to work.

After a long day of work in the dental clinic, Emma went to classes at the university. She found her classmates talking about how they could help her. Since her greatest wish was to return to her country and start a foundation to help and educate abandoned and abused women and children, they tried to come up with ways to send her back with sufficient funds and protection. However, Emma's greatest fear was about being accepted in a society where women are not valued.

As they saw Emma coming, they asked her how she was. Emma's

eyes were puffy and swollen. She told them she couldn't sleep well. She didn't mention Karim, though, at heart, she was a wreck.

"Emma, we have helped you to make your dreams come true. We will help you with the foundation you wish to raise in your country," said a classmate.

"We have a plan in mind," said another. They shared the plan with Emma, who was speechless.

"Say something," said one of her classmates.

"I do not know what to say," she responded.

"The only thing I can say at this moment is thanked you very much, everyone. There must be a purpose in everything that has been happening to me. Arriving, being supported by Christian missionaries, knowing you all—all of this implies there is someone up there watching over me and my children."

The classmates smiled.

"Now listening to you guys wanting to help me return to my country to raise a foundation for battered women and children, this is God's blessing upon me, although I am afraid." Emma felt like she would break into tears.

"Everything that happens under the sky has a purpose, so do not be afraid, because everything will turn out well," answered one of her classmates. She hugged Emma.

Emma felt at peace. Tears began to roll down her cheeks, but she wiped them away. She didn't tell anyone that her husband had been killed in Afghanistan. For most people, they would tell her to have no remorse. It is all about what the heart wants. It is difficult for people to get over someone they loved. Even if the feeling of love had been buried long ago, a sudden mention of that person could rekindle the fire.

After sharing with her classmates, Emma felt surrounded by people she knew and not strangers.

"My daughter gave me some tickets to invite people to an event at her school, Williams High School, on Friday at seven p.m. I would like you to attend this event with your two children. I'm sure this story will help you a lot. A Spanish teacher will tell a fable. He is better known by the student as Uncle P, and the story is 'Challenging the Unknown,'" said Mansur Kadal, a compatriot of Emma.

Mansur's parents had to leave the country with other family members, including Mansur. They had been threatened for their lives. His father was a rich man and of great influence in the country.

"Thanks for the tickets; I'll be there with my two children in the front row!" Emma replied with a smile.

5

Turning the Pages

Having a different challenge to living up to each day, it is not always the case that we emerge. How we decide is always guided by our experiences.

Decades back, Jacob's father, Aidan, went to the bank to apply for a loan to renovate his restaurant. The bank denied him a loan. The representative said, "We are very sorry, sir, but at this moment, the bank has denied the loan."

This news came to Aidan with a thud. He didn't expect the bank to deny him a loan despite his financial turmoil. "But sir, I depend on this loan," pleaded Aidan.

"Sorry, I cannot help you with this. The bank has to assess each individual for their ability to pay. We do not think you will be even able to pay back the markup. I am very sorry," replied the representative.

"This may lead to my restaurant being shut, sir," replied Aidan.

"But I cannot help you," replied the bank representative as he brought his hand forward to shake hands with Aidan.

People rely on loans as a last resort. Aidan had an outstanding debt. He had fallen behind on payments; the bank had denied him the loan.

Aidan had to try one last time. "Our service has come down a lot in recent months because of the construction of a new shopping center across the street. We are trying our best to tackle new developments in the area. However, we are still suffering losses. I would not want to take another loan, but because of the new restaurants in the area, I have to do it."

"I'm sorry, but I cannot do anything for you or your business. I also want to remind you that if you do not make a payment soon, we will have to take over your property to make up for the loan that you aren't able to pay," replied the bank representative.

However, is it sufficient to gauge everything based on one bad experience of the past? Should we let the birds be free in order for them to learn to fly? Or should we keep them in a cage? Should the fear of failing stop people from pursuing their dreams?

Jacob was a young man who dreamed of having his own chain of restaurants one day. After graduating from college with a business administration degree three years ago, he got a job as the General Manager for a restaurant chain. He was a good student at the university and a competent employee at work, so everyone told him he would go places. Since his childhood, Jacob has dreamed of starting something on his own. He aspired to make a difference and be his own boss.

During the three years as the General Manager, he presented a small model of work and innovation that could make the restaurant chain grow thirty percent in revenue and customers. This motivated him to be interested in opening his own restaurant. Although, whenever this idea is brought into consideration, he feels a little insecure. This is because his father had his own restaurant and after ten years in the business, he had to close it due to a fall in revenue and declare bankruptcy.

Jacob consulted his father and presented him the idea to opening his own restaurant. His father disapproved of this idea.

"I don't find this plausible, my son; you shouldn't go ahead," said his father.

He had his own fears. Being a parent, he doesn't want his child to suffer. After ten years in the market, his business went bankrupt, and he did not want his son to go through the same experience.

"But why dad? This is a good idea, a complete plan, and something that only needs to be put in action."

"You don't understand; you are too young for it. You want to be in charge, but you are not ready for the risk that comes along with that responsibility," replied Aidan.

"But, Dad, let me explain."

"Please don't tell me you are ready for that grind. Tell me! Can you see yourself losing everything overnight? Are you ready for this?" replied Aidan with disagreement.

"Why would I make mistakes like that? I wouldn't follow you. I have everything planned, Dad. Why do you think I will become the same example of a failure as you are?" Jacob blurted out.

There was a moment of silence. Aidan said nothing. He was short on words. He opened his mouth, but nothing came out. His eyes began to fill with tears.

Failures leave a big and adverse impact on a person's mind. Can people grow out of it? Yes, they can, but no one can predict when. Aidan is an example of a human being who had suffered failure. He hadn't grown out of it.

When he tried to stop his son from paving ways for his own startup, it was not because he doesn't trust him, but it is his fear speaking.

"Dad ..."

"Please, Jacob, no more." Aidan didn't want to hear any more.

"Please, let me explain."

Aidan didn't say a word.

Jacob tried to explain the restaurant model he wanted to create, comparing it to the model his father had. His father kept insisting and didn't even spare a few minutes of his time to listen to what his son had.

"Dad, please say something. I know I have been unreasonable, forgive me."

After some time, Aidan spoke. "What guarantees you that your restaurant will make a difference?" His father was dismal about the fact that Jacob had been pressing his idea and bringing it up again and again.

Jacob became afraid his business would not work and he would have to close it and declare bankruptcy, just like his father. Even then, this young man remained unstoppable. This did not make him give up; even when his father was not willing to support him, he had set out to make his dream come true. All by himself.

On Monday, during work, Jacob went out to buy lunch. He met a friend named Justin. They'd become friends two years back and had lost contact.

"Justin, that's you!"

"Jacob, it's good to see you; how are you?"

"Not very good," replied Jacob.

"Why is that so?"

"I need someone to let out everything that has been poisoning me."

"Let us sit somewhere and talk," replied Justin.

Jacob told him the entire scenario.

Justin understood Jacob was dwindling with what he had done to his father, and that he still wasn't allowed to pursue his dreams.

It was good to meet someone who cared about Jacob so much.

Justin's son attended Williams High School, the same school where Uncle P would read his story, "Challenging the Unknown."

After meeting and having lunch together, his friend gave him a little more information about the school and invited him to the show. He handed him two tickets. "It will serve as a learning experience!" said Justin.

"Take your father along and ask for forgiveness. I know it will work," said Justin.

He gave Jacob tickets for Uncle P event.

Jacob was not sure that he could attend the event, since his work schedule made him travel, sometimes unanticipated.

On Tuesday, Jacob invited his father to have breakfast outside, and during the breakfast they talked.

Jacob told him about being invited to the school. "A friend invited me to an event in a high school. I have two tickets. I want you to come with me."

"I will think about it and let you know," replied his father.

Before getting out of his chair, Aidan advised his son Jacob not to get into debt and to stay working in the restaurant chain. According to him, that was better.

"My son, it's not that I want to discourage you, but I think it will serve you better in a full-time job. The gastronomy industry is a great competition, and only those who have a lot of money to invest can stay afloat. The voice of the experience tells you, 'Look at what I had to go through, after ten years in the industry, look at how I finished everything.' I advise you to play it safe. All the ideas and plans you have to innovate the industry, present it to them; they have the capital to invest, and you do not. Don't make yourself endure the pain," said Aidan.

"Father, please, do not bring that to the table. Who knows if that will not work with me? The industry is good. Who knows if I become a big fish in this field," replied Jacob. "Do you know what your mistakes were?"

"That you do not listen to anyone. I know that I do not have the experience you have in the gastronomy industry, but if you had at least listened, things would have been much better. I told you several times that you should renovate the restaurant when you still had the capital to do it; this was because the restaurant was doing well. But you got stuck amid the chaos in the industry. This was during the time while everyone in the area was renewed and making changes to bring more customers to their businesses."

"Have you invited me to this place to accuse me?" replied Aidan.

"You are taking me wrong dad."

"I need no more truth slaps and slaps of conscience about me. With the experiences of the past; I think I have had enough. And if I failed, what guarantees that you will be better off?"

Aidan cut him off. "You say that I don't listen to anyone; you don't either. Before you accuse me, think about what you're doing at this point. Another thing, because you studied business administration, and you have progressed in recent years and you are doing well, that does not give you the right to disrespect me. Who paid for your studies? Tell me?"

Jacob's father got up from the table and left.

A few hours later, Jacob called his father just to make things right, but he didn't pick up the call.

The next day, which was a Wednesday, after work, Jacob returned home and tried to call his father during the day and night before going to sleep, but again, he did not answer.

Jacob left a voicemail, asking for forgiveness for the way he had addressed him. "Hello Dad! I hope you are well. I know you're upset with me and you have reason to be. I should not have addressed you in that way; forgive me. You are very important in my life and if I come to you; it is because your advice mean a lot to me. I know you do not want me to do in business and you want the best for me, but how can I go against my will? If this passion that I carry inside consumes me every day, why are you stopping me? We are always just a decision away from a different life. You do not know how many times I have wanted to give up and abandon this dream and do what you have told me, but every time I try, it always comes to my mind and my heart pounds. Sometimes I have spent nights in candlelight weighing if God created me for this, because I enjoy my work and I like to have people contact me with their feedback on how to improve the service. I don't know if this will suffice to you, but that is all I have to say. You are very important to me. Please speak to me, at least once."

Jacob hung up the phone. His father saw the phone ring, but he did not answer it because he was upset with his son. He then listened to the voice mail. He still didn't reply.

Being a witness to the fights between Jacob and Aidan, Jacob's mom left him a text saying, "Come over tomorrow in the afternoon and talk to your father."

The next day, Jacob visited his father around four-thirty. He rang the bell. His father approached the door. When he opened, he realized his son was standing in front of the door.

Jacob noticed his father was watching a TV show. His father invited him in. They had a conversation where Jacob apologized to his father for his unruly behavior again.

After a while, they reconciled and his father realized he cannot stop him from going against his dreams and asked for his forgiveness on the fact that he tried to persuade him to surrender his dreams, owing to the fear of failure.

Jacob told him that on Monday he had a meeting at three p.m. for a loan with the bank representatives. He also invited him to go out to dinner, and then to accompany him to the school event, where they will listen to the story of Uncle P: Challenging the Unknown.

His father agreed to go out with him and accompany him to the event. In this way, the father and son duo got to spend some time with each other.

It is often seen that we let our fears overcome our will have. How will Aidan and Jacob go about it? It is yet to be seen.

**

Ariel was in a deep sleep. A depressed soul sleeps more, but can this be helpful? They always have to come back to the reality; they always have to wake up in the world where nothing has settled.

Ariel, a sixteen-year-old boy, was going through the same. It had been hours and his parents hadn't heard from him.

"Can you please check on Ariel?" asked his father.

"Wait, let me see," replied Jessica as she paced her way toward Ariel's room. She opened the door of his room. Ariel was in a deep sleep. Jessica went inside. She sat close to Ariel. She could see the traces of tear drops on his face. He must have had a hard time. It was clear.

Jessica planted a kiss on her son's cheek and said, "Why do you have to penalize yourself, my son? Why have you lost hope? Seeing you like this is heart-wrenching."

Ariel didn't reply. She then got up, put a blanket on her son, and left the room. She closed the door, ensuring she made no noise.

As Jessica came back to the living room where Peter was sitting, she had a tensed and perplexing expression on her face.

"Is he okay?" asked Peter.

"No, he isn't. Our son is in immense pain. He has been humiliated, and some teammate has held him responsible for something that happened in just a flick of a second. Why does it have to be like this for him? The time when he needed some appreciation and recognition—look at how life has treated him."

"Hey, hey, my dear. I can see the grieved and concerned mother speaking. I know you cannot see your son suffer. I promise that you will soon see your son smiling." Peter rose from the sofa and kissed Jessica on her forehead. He could see that Jessica was not at peace.

"Please smile, my dear. If you act like this, how do you think Ariel will act? He needs our support and you, his mother, are his support."

To this, Jessica nodded and said, "I will make sure I keep my calm in front of him, but I want to see my son smiling soon."

"I will make sure he does."

After sleeping for hours, Ariel woke up at six pm. He had no intention of leaving the room. On Jessica's continued insistence and pursuance, he agreed to leave his room. He got up and went to the bathroom to take a shower. As he took a shower, he could hear every word of insult in his head. It was coming like a flashback. Some teammates, laughing at him, mocking him, and holding him responsible, tears fell down his face, and he held his hands on his ears. Did they stop the echoes? No. They didn't.

Ariel came out of the shower, put on his clothes, and left the room. It was already supper time.

As he entered the living room, Jessica addressed him. "My son, sit down. I have prepared dinner, delicious cookies and banana cupcakes for you. I know you them."

To this, Ariel didn't reply. He sat on the sofa and nibbled on the food. He wasn't in a mood to eat. His eyes were swollen.

Peter knew Ariel was sad, but to cheer him up, it was important to make him talk his heart out.

Peter tried to start a conversation. "My son, no one likes to be defeated in any area of his life, but we understand that losing a game does not mean the end of our life. We all are given chances. This was not the last chance. Adversities teaches us that in life we must learn to win and lose; and when we lose or fail in our attempt, that shows us that not everything we want to achieve in life will be easy. Everything will cost an extra effort. Never forget, failing only teaches us the worth of winning. It encourages us to make sure we win the next time, why look on the negative side? Look on the positive side."

To this, Ariel didn't reply. He was silent. He took longer than usual to chew the food, so he didn't have to talk.

Jessica sat by Peter's side. She had her hands folded and looked at Peter with an expression of, "Please try once more."

Peter once again tried to talk, "Just because you weren't able to drag your team to the semifinals single-handedly, and some schoolmates have bothered you, it doesn't decide your fate. That is just one part of the story. How can you know if the most famous basketball players were always as good as they are now? Don't you believe in learning? Don't you believe in falling down and getting up again?"

To this, Arial blurted out, "Dad, they insulted me in front of everyone. They called me incompetent. I was called a loser. It is hard for me to get over it."

Then Peter said, "Many times, the surrounding people will be the first to doubt our abilities and potential. They themselves do not have it, and they want to use those moments when things do not go well to attack us and use it against us. People try to put us down whenever they have the chance. Do you know why they do it? Because they can never be like you. I know that next season will be much better."

"I don't want to take part in basketball; not anymore."

Jessica raised her eyebrows in concern.

Arial continued. "I don't know if I will ever change my decision again. My teammates told me I was a loser. They told me that basketball is not for me. Maybe they are right! Maybe I should try something else. I am not afraid of them, but I am afraid of being treated this way ever again. What if they were right?"

Peter replied, "You know, this is the catch. You need to overcome your fear! Hold strong during the adversities. No one is born an expert. We can, however, become one."

To this, Ariel didn't reply. Peter tried to cheer up Ariel by telling him an entertaining story. "Ladies and gentlemen, I want to introduce you to Ariel, the most valuable player of the year. Now he will tell us his experience of how he came to be a professional basketball player."

Then he imitated Ariel and said, "Thank you for this warm introduction. It is good to be here tonight! I feel honored! I would say, it isn't easy, but also, it is not impossible. It takes time, effort, and dedication! Many people around me didn't believe in my capabilities. I was called out names, I was mocked, and I was made fun of. People told me I was wasting my time because many people around me did much better, but I didn't pay heed to such comments. If I had done so, I wouldn't have been where I am today. I want to tell all of you that do not be demotivated from what people say, you know your capabilities and know that you can do it! Achieving anything requires persistence, discipline, and dedication. If you have that in you, you are deemed to succeed."

The 'supposed' Ariel continued, "I would like to thank everyone who supported me throughout this journey. Without all of you, I wouldn't have made it through. For those who stood by my side, you can see, here I am today. I owe my success to all of you."

In imitation, Peter then became the entire audience and clapped for Ariel. Jessica too began to clap. On this, Ariel began to smile and laughed. Seeing their son laugh, Jessica and Peter's face brightened.

"My son, tell me now, are you going to practice on Monday afternoon?" asked Peter once again.

"Okay, we will go," replied Ariel.

Peter jumped on his feet and began to dance. To this, Ariel and Jessica began to laugh again.

Peter halted and then said, "My dear, do not worry! Everything will fall back to place; trust me!"

Ariel got up and held Peter's hand with a smile on his face. "Thank you for cheering me up, Dad!"

"That's the job of every good father who wants his children to achieve new milestones in their lives. I always want to be there for you. I will do everything in my power to make sure that you always get the best. I will always clasp your hand. I will never let you fall. For me, it is an honor that you are my son and I am grateful for that. I love you!"

"I love you too, Dad. You are the best." With this, Ariel hugged Peter.

On a Monday afternoon, Peter came home from work and saw Ariel set for the practice and waiting for him on the sofa. He changed his clothes, and the family went out for practice.

At the park, Jessica sat under the tree on the grass and Ariel and his father ran two laps to warm up and then started practicing shots on the hoop.

After a few minutes, they took a break for water before they moved further in practice. Jessica passed two water bottles to Ariel and Peter. It was a sunny afternoon, and she wanted them to stay hydrated.

Peter passed the first water bottle to Ariel, saying, "Here, here. My future basketball champion should drink water first. Take it, champ!"

Ariel took the water bottle with a faint smile. No matter how much he appreciated his parents' efforts, he still couldn't get his mind off the fact that he had failed in the last game. His heart was pounding, and the thing that stuck in his mind was, What if I let these people down? What if I don't have the potential and they are trusting me too much?

Ariel drank a fair amount of water, wiped the sweat from his forehead, and told himself in a silent whisper, "You can do it."

"Did you say something?" asked Jessica.

"No, Mom, I'm just waiting for Dad so we can go back for practice."

"I am happy that you've made a comeback. You're the real champ, my son! I'm sure you'll go far in your life. You'll see a lot of success in your life. Don't let the negative words get to your mind; just shut the bad words off. How can anyone else know your true potential other than you yourself? Don't forget: we're here to support you through and through. I know you'll make us proud."

After listening to what Jessica said, Ariel was determined to live up to what his mother expected of him. As Peter came back, the son and father went back for practice.

It was a good practice. On their way back, Jessica said that there was no milk in the house. For this, Peter took them to the supermarket.

"Ariel, do you want something from the supermarket?" asked Peter.

"No, thank you, Dad!" replied Ariel.

Ariel waited in the car while Jessica and Peter went inside the supermarket to fetch a few things, including milk. In the supermarket, Jessica took out a few tickets for the Friday show and handed them to the people shopping. As promised to the volunteer, Jessica distributed the tickets and kept three aside for the three of them. She also bought ice cream for the family.

Jessica and Peter came out of the supermarket and sat in the car, and then the family went home. Jessica prepared dinner for the family and this time, Ariel came to the table without her or Peter having to call him.

Ariel looked fresh and happy. His face was content. Jessica and Peter both were at peace to see their son happy.

After dinner, Ariel waited for his parents to finish. Before everyone left the table, he said, "Mom, Dad, I want to say something."

"Yes, dear, go ahead!" replied Jessica.

"I already feel much better. Thank you for supporting me at all times. I will owe you both for the rest of my life. I have the best parents in the entire world. I do not know what I would have done without you."

"You can always count on us, my child. You are our only child, and we wish the best for you. We want to see you succeed and we want to see your dreams turn into reality," said Peter.

"You have been a blessing for us. We are proud to be your parents. You are the engine that moves us. What would we do without you? Our lives would have been colorless," said Jessica.

Ariel got up and hugged both his parents.

"Whenever I see you, my son, I think about the first time that I held you in my arms. So many years have passed. You would cry if I put you down. When you were three, you would make innocent faces just so I would tell you one more story before you go to bed, and look—today, you are on your feet, pursuing your passion. Though the fact that I still tell you stories hasn't changed a bit," said Peter.

At this, Ariel began to smile. It was a good day for all three of them. "You can always count on us, my child!" said Jessica.

"I know Mom, and I am the luckiest child on the face of this earth!" replied Ariel. With this, Ariel asked for his parents' permission to go to his room. Both Jessica and Peter planted kisses on his cheek, and he was off to bed.

Things had fallen back into place for Ariel. He was happy and coming to terms with the fact that he could make a comeback!

Ariel and Peter had practiced basketball on alternate days.

"Wait, I need to grab some water." With this, Ariel went to Jessica to get his water bottle. Jessica, who sat in her regular spot, handed him the water bottle. Ariel sat and began to sip. While sipping, he took out his phone and began to check his text messages. As he checked through his messages, he saw Daniel's messages. He opened the message, and what he read left his heart pounding. The text read, "Hello, Ariel, I hope you are well. Even after being told off by the coach, Albert and Joel have continued to speak against you. Check the link attached."

Ariel began to sweat even before opening the link. He held himself strong and tapped it. Joel and Albert had been bullying Ariel in a video they uploaded to social media.

Albert said, "Because of that one pathetic player, we couldn't qualify for the semifinals. He is a loser. He should abandon basketball. I just have one message for you, Ariel: surrender. Do not waste your time."

Wrapping it up, Joel said, "They consider him the best player. He's just a loser. He can do no good for the team."

Ariel had tears in his eyes. He was like a mentor to his peers. He always liked to spend hours with Albert and Joel and other teammates in practice. He kept nothing from the past in his mind. He had looked forward and let the past go, and how did they pay him back? By trashing his image in public.

He kept his phone in his pocket and sat by Jessica. Peter, who saw him not coming back, asked him to come back. To this, Ariel replied, "I don't want to keep practicing. Let's go home. I think we are wasting our time."

To this, Peter and Jessica became concerned. Peter asked, "You were fine a few minutes ago. What made you change your mind so fast? Let me see your cell phone."

"No, please. Let's not take it there," replied Ariel.

"I HAVE TO SEE YOUR PHONE!" replied Peter authoritatively, and he took the phone from Ariel. He saw the link and watched the video with Jessica. Then he locked the phone and looked at Ariel. He had tears in his eyes.

"Basketball may not be for me. Everyone at school sees me as a loser. Why don't you transfer me to another school? I don't want to go back to the same school next school year." Ariel began to cry.

"Why, my dear? Didn't I tell you that you shouldn't let such things get to your head?" said Jessica.

"You don't understand. I feel like everybody has made fun of me so many times."

"Ariel, if you say so. Let's consider your proposal for once. But let me tell you a few things," replied Peter.

Tears rolled down Ariel's cheeks.

"You stop crying. You are not a little boy any more. Now, listen to me."

Peter held his hands, he then continued, "Let's go by step, my son. I have three for you. First one: 'Everybody' is too many people. I do not think everybody in your school considers you a loser. There are only some of your teammates who, out of jealousy, have started a defamation campaign with the purpose of hiding behind their attacks. The second step is, changing schools will not solve this situation. Never! You are letting your enemies know what they are saying is true. The reality is far from this. There will be times in life when you will have to face such people who will work their best to knock you down because they want to stop your success. Why is this so? Because they can never be you! But you have the final decision; you prove that you are right or that you are wrong. It's up to you. The ball is in your court!"

Ariel didn't reply; he was not buying what his father said, and how could he? For a teenager, is it easy to take such defamation attempts? Not that he was scared; he was heartbroken.

"Do not let what others think or say about you affect you. I have said this a lot of times, and I will say it once again: these are tactics of a mental game to weaken you. They hate you because they are not you. Never forget this. Who has ever made it to team captain at your age? Who has ever gotten the chance to train his peers as you did? Tell me," said Jessica.

Ariel stayed silent. He was weeping.

"For many, you are an example to follow, for others, a challenge on the road. They are trying to deal with you in their way. For the first time in five years, the Jefferson basketball team reached a semi-final. And, do you know what the important part is? That those who are attacking you were there from the first day the team started and could achieve nothing to make the team advance. It was you who steered the sinking ship out of the storm. They are older than you and yet they had to be trained under you. Why don't you look at things in this way?" asked Peter.

"You guys are trying to cheer me up, but it is not working. What if everyone thinks the same as Albert and Joel do?" replied Ariel.

Jessica took a deep breath and said, "Why did so many people come to watch the game? There was a reason: for the first time, many people went to the game to support the Jaguars at Jefferson's basketball court. Remember, these words: a single person can make a difference in the lives of other people—be it a school, a community, or in your case, a basketball team. Do not give up, because your story is just beginning and only you will determine the path it should take; there will be many obstacles and opposition along the way and, it will not be easy, but neither will it be impossible. It is wise to give up? Tell me?"

"I appreciate your words, Mom and Dad. What Albert and Joel said struck me in the heart. My heart aches to see how they have been talking about me."

"Every decision will cost you something; never forget this. There will always be someone who would die to win wars against you. You may have never wronged them, but they will still wage wars against you. If you want to let them hit you, then make sure you are armored," replied Peter.

"Yes, you are right! But, could we continue the practice another day," said Ariel as he wiped his tears.

"Okay, then pack up, we will head home," said Peter.

"But Dad, what was the third step?"

"The third step is that I will call the coach so we can have a meeting with the parents of those two young boys. We cannot ignore this form of attack; they must stop it before a tragedy occurs."

6

The Lovers Reunite

On the following Wednesday, Jeffrey picked up Amanda at seven-thirty p.m. Amanda wore a beautiful black dress while Jeffrey was in a classy tuxedo.

As they drove to the restaurant, Amanda shared her tales of New York and Jeffrey was more than interested to know about it. As they reached the restaurant, the waiter guided Amanda and Jeffrey to their reserved table.

"Oh, I have not visited this place for quite some time. The place has changed much, including the interiors. Also, this place has been expanded," said Amanda.

"Yes, I'd say that the place needed it, we all need to embrace changes when the time calls for it," replied Jeffrey. To this, Amanda smiled.

"The place looks beautiful, however," says Amanda.

"Just like you," says Jeffrey in a slow whisper.

"What did you say?"

"Um, nothing. Was just agreeing to what you said," replied Jeffrey.

"If you say so, okay," said Amanda.

Jeffrey let out a sigh of relief. Amanda had nearly caught him.

As divine help, the waiter intervened to take their orders, and Amanda forgot about this.

Amanda and Jeffrey began to talk about different things until the waiter returned with drinks.

Both of them sipped from their glasses and then Amanda said, "You know what, during our first interview, your English accent had made me nervous beyond measures."

"I don't think so."

"Aha, why?" replied Amanda in intrigue.

"It must have been the journalist," Jeffrey replied with a grin.

"Maybe."

"Well, well, Amanda, you know why I was so eager to meet you?" asked Jeffrey.

"I would want to know from you," smiled Amanda.

"Look, from my experience as a journalist and from getting to know many people's stories, I know one thing. We should never hesitate or wait in acknowledging and sharing our feelings. This life is too short to wait. No one knows when we will take our last breaths."

"Oh God, what happened, Jeffrey?" Amanda became tense.

"Amanda, I am sorry to bring this topic, but I have done all my waiting. Enough time has already passed to decide in your life. You are a beautiful woman, with all the characteristics that every good man wants to find in a woman. Have not you thought about starting over?"

"Um, Jeffrey, let us not go there!" Amanda blurted out.

"No, we need to. I want an answer. Why are you doing this to yourself? Do you not feel worthy of being loved?" impressed Jeffrey.

To this, Amanda opened her mouth, but nothing came out. Like her mouth had dried, or maybe she too began to contemplate.

"Say something, Amanda," Jeffrey brought Amanda's attention back to the topic.

"Jeffrey, my beloved Jeffrey, I will have trouble answering this question. If I was so worthy of being loved, then why wasn't I loved by my husband? Why did he leave me? At this stage of my life, I feel like it will be difficult to start over. I cannot prepare myself to endure the same pain once again. What if I was the one who couldn't keep a relationship? What if I was the one who caused the doom for my twenty-five years of married life?"

"But, your ex-husband cheated, Amanda. How can this be ever your fault?"

"See, he cheated, I know! But maybe I wasn't able to love him enough, or I wasn't a homemaker."

"Amanda, please."

"You tell me, how do I know that the same would not be repeated? I cannot suffer a heartbreak once again, because this time, I could not gather my broken pieces."

"You cannot think like this. Every human being in this world may live a happy life and so do you! It is only up to you to give yourself a chance, and you should!" said Jeffrey in a persuasive tone.

"Jeffrey, look, even if I try to give a chance to someone in my life, getting to know them will take a lot of time, and I don't have that kind of time to invest in right now. I know how most men think, they don't want long-term commitments," said Amanda.

"But not every other man is like that. There can be men who want you," said Jeffrey in a low voice.

"Why would a man ever want me? My beauty is fading. Look at me; my face already has wrinkles. You tell me! Do you want me? Do you feel attracted to me?" asked Amanda.

"I …" this sudden question made Jeffrey nervous and speechless.

"Say it. Please, Jeffrey, don't hold back. I am open to any kind of answer," impressed Amanda.

Before Jeffrey could say a word, the waiter came with the order. Both Amanda and Jeffrey finished their dinner in silence. Jeffrey tried his best not to talk.

Amanda would not let this thing go so. Soon after they were done with the dinner, she said, "So you haven't answered my question yet."

"Amanda, I don't want my honest answer to ruin our friendship. You should understand that you are precious to me. I cannot lose you because of any stupid confession," said Jeffrey.

"Stupid confession? Can you not give me an answer? I am a fifty-five-year-old woman, and I have gone through very difficult times in my life. Nothing can break me like this; go ahead," assured Amanda.

On this, Jeffrey smiled and said, "Look, I like you a lot. I like you so very much! I never confessed this for fear of losing you as a friend. The first time I interviewed you, I wanted to ask you for a date soon after the interview had ended, but the words just didn't come out; I knew that woman like you would have never agreed."

"You are so sure about me and so unsure about yourself," said Amanda as she continued to laugh.

"Then tell me, what would have been your stance?" asked Jeffrey.

"See, I admire you a lot. I feel alive when I am around you. I am at my happiest. I am carefree, I am satisfied, and I am living. This is how I can define your presence for me. I wouldn't have waited to tell you this, but I am still afraid to start over, even if it is with you."

"Give it time, Amanda. I am happy that you feel the same. Let things be. Everything will fall into its place."

"How?"

"I will tell you how: just make one commitment with me."

"What commitment?" asked Amanda in desperation.

"Just keep spending quality time with me."

"I would love to, Jeffrey!"

"Then we are going to a show this Friday. It is a show by a Spanish teacher known as Uncle P. and you cannot say no!"

"I will accompany you!" replied Amanda with a smile on her face.

With this, both of them smiled and savored the delicious dessert before they called it a night.

***

It was a Monday morning and a bright sunny day! At nine in the morning, Oliver received a phone call.

"Hello?" said Oliver as he picked up the call.

"My son! How are you?" replied the woman on the call. It was Betty, his mother.

"Mom, how are you? I am doing well," replied Oliver with great energy. He was happy to hear his mother.

"Not well," said Betty.

"Why? What happened? Did you call the doc?" Oliver was puzzled.

"Because I haven't seen my kids for so long, I want to come to your place to meet you, Anthony, and Leslie. Are you guys at home?"

"Mom, you freaked me out," replied Oliver.

On this, Betty began to laugh. "My son, I was just teasing you, and you should be teased; you do not miss your mom much. This is my way of showing my importance," said Betty.

"This is not the case, Mom. You are most important and precious to me. Now, tell me when you plan to come."

"Right now," replied Betty in a warm tone.

"Sure, we are all at home."

"See you then!" Betty hung up.

At that point, Sofia, who was trying her best to reach Mr. Anderson's place on time, was stuck near a construction site because of a huge influx of traffic. This was because all other roads were under construction. The time that she spent in the traffic made her anxious. She knew she would be late and thus called Mr. Anderson to inform him beforehand.

As soon as Mr. Anderson picked up the call, she blurted out, "I am very sorry, Mr. Anderson, but I think I will be running late today. I'm stuck in traffic."

"That's no problem, Sofia. Take your time."

"Oh, also, how are you?" replied Sofia in nervousness. She had started off with no greetings and thus was trying to make up for it.

"I am doing well, Sofia. Please drive with care; we will talk once you get here," replied Mr. Anderson. He was smiling on the phone at how Sofia tried to cover the situation.

"Thank you so much, Mr. Anderson. See you!"

Betty reached Oliver's place in about ten minutes. Oliver heard the honk of a car. The main gate opened, and the car parked inside. As Betty stepped out of the car, Oliver, standing on the balcony in the second floor facing the garden, saw his kids running toward her. Betty hugged Leslie and Anthony and the three of them entered the house. Oliver went downstairs to greet his mom. She hugged Oliver.

"Oh my son, I was longing to see you," said Betty.

"So was I, my dear mother," replied Oliver.

With this, Oliver took her mother to the kitchen where he was trying to prepare breakfast for himself and the kids. Betty sat on a seat by the kitchen counter and saw her son struggling to prepare sandwiches for breakfast. She stood up, washed her hands, took the knife and the bread slice from Oliver's hands, and began to prepare the sandwiches.

While she prepared the sandwiches, she said, "It's about time that you bring a wife for yourself so your old mother wouldn't have to drop by to prepare sandwiches for you, my dear."

"Mother, please, let's not talks about this in the morning," replied Oliver.

"Listen, I will keep on bringing this until you agree to what I am saying. Look at yourself! Why do you have to do all of this by yourself? You deserve happiness and companionship! What are you waiting for? Why do you not give yourself another chance?"

"I have many responsibilities, and I do not have time to look for a wife. My children are growing, and I want to dedicate time to them," replied Oliver.

"This is how you will justify yourself?" Betty replied. "Why don't you bring up an excuse which has the potential to hold? The nanny is the one who is taking care of them. Listen to me, Oliver, Anthony, and Leslie need to see a female figure in the house; they too deserve to grow up in a complete family environment. Once your kids grow up, they will become busy in their lives, and what will you do then?" asked Betty.

"Mom ..."

"Listen, I haven't finished yet." This time, Betty became furious. She was like any other mother who couldn't see her child suffers. She continued, "How long are you going to stay like this? Why are you penalizing yourself? How will Leslie know what to look for in her love if she doesn't see that in her house? And what will Anthony learn? How will he know how to treat a woman right once he gets married? These kids are young and if you get a woman who understands and loves them, and also loves you, your life will change for the better."

Oliver didn't reply.

"Son, I know you will prove to be the best husband, you already were, but your wife is gone to heaven, and she will never return. Why do you want to deprive this house of a woman, yourself of love, and your kids of a mother?"

"Why are you so worried about me, Mother? I am already loved enough by you and by my children. I need no more. I am an adult. I can also take care of myself."

"Yes, I can see how you could prepare your breakfast."

Betty held his hands and said, "You still haven't told me why you do not want to give yourself a chance. Tell me, so I can free you of your insecurities, my child."

"Mother, I'm afraid to start a new relationship. In the position that I am, it will be difficult to find someone who can love me for who I am and not for what I have. I know many people who would fall for my money rather than me. How can I trust anyone in such a world? Apart from that, I would not like to see someone mistreating my children. You know that they are the treasure my deceased wife left for me, and I cannot see them hurt. If someone wrongs them, will she ever forgive me? To find someone who treats them as their mother did and who has a love for my children will be a difficult task to complete. I hope that you understand me now."

At that moment, Sofia opened the door that was already unlocked. The children ran toward her, embraced her, and kissed her.

"Sofia, we missed you," squealed Leslie.

"I thought you wouldn't come today," added Anthony.

"Why wouldn't I come to meet you both? I am always looking forward to Mondays," replied Sofia. With this, Sofia embraced the kids once again.

The kids said, "We love you, Sofia."

"Sofia loves you more," replied Sofia.

Betty and Oliver saw and heard everything from the kitchen. Then his mother looked at him and smiled.

"Now, what is this supposed to mean, Mother?"

"The woman who will love you and your children is already in front of you. Someone who would never wrong them and someone who isn't after your money; she has been there for so long, in front of your eyes, you had just chosen to ignore," replied Betty.

"Mother, she is the ..."

Betty cut him off and said, "So? What is wrong in that? My son, happiness has already knocked on your door, and you only have to choose one of the two options. Either you let it in, or you leave it outside. You choose."

Before Oliver could reply, Sofia approached them and greeted Betty and Oliver.

"I am very sorry for being late," said Sofia.

"Oh, that is just fine, sweetheart," replied Betty.

"Sofia, why do you worry so much? It is fine," said Oliver.

"I feel that I have been making mistakes lately."

"No, you haven't," replied Oliver.

"I have also been rude regarding the conversation last Saturday. I apologize that I did not answer your question. The thing is, I felt somewhat uncomfortable answering your question, but I shouldn't have avoided your question like that. I am sorry," pleaded Sofia.

"Sofia, I didn't mind; that question was personal. You do not have to apologize," replied Oliver.

Then, Betty, eager to know what the conversation was about, intervened and said, "Why do we not meet at my house this afternoon? I will prepare the dinner early and we all can sit down at the table and talk. What do you both think?"

Oliver looked at his mother with a disapproving expression. Then he said, "What do Anthony and Leslie say?"

They were excited to go to dinner at Grandma's house along with Sofia. "Yes, yes, and yes!" both kids said.

"I'm willing; what does Sofia say?" said Oliver.

"I do not know. I wanted to get home early this afternoon to organize some things," she replied.

"Sofia, you cannot disagree; please join us!" said Leslie.

"Yes, please," impressed Anthony.

"Um, okay," replied Sofia.

The children gave a shout of joy. "Yeah! It will be fun!" said Anthony.

"Then the afternoon affair is completed. I need to leave for work. I will see you all in the afternoon," said Oliver.

"I also need to take the kids to the amusement park; see you guys in the afternoon." With this, Sofia took the kids and headed to her car. Betty bid goodbye to Sofia and the kids and told Oliver to stay. They talked about this development.

***

After having a fun-filled day at the amusement park, Sofia and the kids reached Betty's house. The dinner was ready. The aromatic smell filled the kitchen.

"It must be very delicious," said Leslie.

"Yes, my dear," replied Sofia.

"I want it! Please, please," squealed Anthony.

"Wait, my dear; let us all sit on the dinner table. Let your father come," replied Sofia.

"I am here," said a voice. Oliver had come in.

"Dad! Come, let's eat, let's eat!" said Leslie as she made a run toward him.

"It feels hot in here. I will adjust the temperature," said Oliver.

All of them sat on the dinner table and began to eat. After the dinner, Anthony and Leslie went to watch cartoons while Betty, Oliver, and Sofia talked in the dining room.

"So how long have you been in Maryland State?" asked Betty.

"Fifteen years. I left the Dominican Republic when I was twenty years old," replied Sofia.

"I would have liked to visit the Dominican Republic; some friends told me it's beautiful. Did you ever return to the Dominican

Republic? Or do you plan to go there? Maybe to visit or to move?" asked Betty.

"No, I have not returned, but I am in contact with some members of my family. They tell me that much has changed since I left; also, I could not recognize most places when I went there. I want to visit them, and I miss them a lot, but I would not wish to move. I want to embrace my family members, but I cannot leave the country."

"Oh dear! Why is that so?" asked Betty with concern.

"I have a work permit right now. I submitted the documents to apply for the green card ten years ago, and I am still waiting for the immigration department to update me on how my process is coming about. If I leave, I can never come back," replied Sofia in a low voice.

"Who else in your family lives here?" asked Betty.

"I have my father who lives in New York and two brothers who are younger than me; they have their own families and they live in New Jersey."

"And your mother?"

"She died when I was ten years old. My brothers and I grew up with our grandparents. I feel sorry that I could not spend much time with my mother. She was taken away too young. Her name was Christina. She was exquisite. After the death of my mother, my father immigrated to the United States in search of a better future for his children," replied Sofia.

Oliver and Betty could see tears in her eyes. She had been left alone and stranded in this world. This was a moment of silence and great unease.

"I am sorry to hear this," said Oliver to break the silence.

"I am sorry, dear," added Betty.

"Two-and-a-half years taking care of my children and I had never sat downed and talk to you about your life. I never knew about the struggle story you have. You are one strong woman. I want to ask for an apology for that. I shouldn't have been so naïve," added Oliver.

"Mr. Anderson, you do not have to do it. I understand. You have a busy life. We all have a busy life. I think sometimes we become so busy in the affairs of our lives we forget to know about the surrounding people. We forget that in this world, everything is temporary; one day we will leave and only the memories will remain—the memories of those people who care about us, we think we will get many chances to make up for the lost time with people who care about us, but the reality is far from this. The time is running away and we don't know when our time is up. The clock is ticking, however," said Sofia.

Oliver and Betty heard every bit of what Sofia said in silence.

"You are right," said Oliver in a low tone.

"Losing a loved one is painful; all three of us have been through this pain. It is hard to let go, and the memories, they stay forever!" said Betty. "But you are young, and you need to live your life. Tell me, have you never gotten married or thought of having a family?"

"This is something that rekindles my sorrows and happiness," said Sofia as a teardrop fell down her face.

"If you don't want to tell, then it's fine; we do not mean to hurt you," interrupted Oliver.

"It is time that I stop hiding this from people, not because I cannot give it away, but I feel like I am living it once again. But I will tell this to you both. Yes, once, I was married. I married the love of my life. It was like a dream coming true."

Oliver and Betty listened with their undivided attention.

"There are situations in life that occur to human beings and we do not know the reason behind those adversities. Sometimes we have no answers to such things, and we always ask ourselves that one painful question: 'Why me?' When my family came from the Dominican Republic, my brothers and I were living in New York with our father. There, in that beautiful city, I met my husband. His name was Carlos. After a time of getting to know each other, we got married. I was in love with him and so was he. He was the best thing that had happened to me. After losing my mother, he was the only one who could provide me with solace. Not that my father didn't take care of me; he did more than he could, but Carlos was different. After several months of marriage, my husband received a job offer as a construction contractor in the state of Maryland. I was happy for him and we moved here. We established ourselves in the state. As a happy couple, our next priority was to have at least two children. We tried our best; we both wanted children, but we didn't see any results. We visited a doctor; we underwent several tests. The results of my husband went well; he was fit to have kids. However, the doctor found some anomalies in my system. He put me on treatments for about six months, but even after that, the doctor concluded I could not have children."

Betty and Oliver looked at each other but didn't say a word. There were many things they didn't know about Sofia.

"It was like Carlos had asked me for just one thing, children, and I couldn't give him that. I was sorry. We visited another doctor to hear a second opinion, and the results were the same. It was devastating for me as a woman, when you want to give children to the man you love so much. I even told him to leave me and marry another woman who would be fertile. But he loved me so much, he found another way! We agreed to adopt a baby. Before completing the process, one afternoon back home, I received a call that tore my world apart. My husband had an accident, where he died on the spot. That accident devastated me. A negligent driver under the influence of alcohol caused the accident. A drunk driver took away all my happiness. The man I loved had left me alone in this world."

Tears ran down Betty's face.

"My world had fallen. I didn't know where to go. I didn't know how to find solace. I couldn't believe I had lost everything in the blink of an eye. I was so eager to have a baby and look at what my life had in store for me! After a while, I decided that if I cannot be a mother, then I can help another mother take care of their children. Thus, I chose this profession. I was working in a childcare for a few months, but the work diminished and I was laid off. I still longed for kids. After Carlos, the only thing that could make me happy was kids. Then, I found a notice in the newspaper where a lawyer was in search of a nanny to look after his children. I never thought I could get this job because of the requirements. I still took the chance and applied, and here I am. This work has been a great experience. Sofia then stopped and began to sip some water. Betty looked at Oliver and smiled, to her surprise, he too smiled back. Before Betty could say anything further, all of them heard Anthony and Leslie fighting over remote control."

On this, Sofia got up to pacify the tension between them. "Please excuse me. I need to look into this matter." With this, she left the table.

She said something to both of them, Betty and Oliver cannot hear, but they see Anthony and Leslie stop fighting.

On this, Betty turns to Oliver and says, "What else do you want? Do you have any other repercussions?"

"Mother, you tell me, how would I go about it? Is it so easy to invite someone for a date?"

"If you do not do it, I am sure that someone else can," replied Betty.

"Oh Mother, what I should do with you?" replied Oliver.

"Do nothing, just act on whatever I say. Sofia is beautiful, sweet, affectionate, and social. She is the woman that you need in your life, you know this, Oliver," replied Betty as she impressed her point.

Then Betty looked up, put her right hand on her chest, and then said, "My dear Elizabeth, you must see your husband from the sky. He needs another woman in his life now, your children need a mother, and I am just trying to help him. I know you will also approve of what I am saying. He loves you with all his heart!"

As her prayer ended, Sofia returned to the table.

"The kids listen to you, don't they?" said Betty.

"These kids are well-mannered; they cause little trouble to me, but I have another thing to tell you guys that I had forgotten about!" replied Sofia.

"What is it?" replied Oliver.

Sofia slid her hand into her bag and took out the tickets she had for the show.

"These are the tickets I picked for a story event called Uncle P: Challenging the Unknown. These are three tickets. Mr. Anderson can go with the kids and have a quality time with them."

"Um, let us do one thing. I can babysit the kids while you and Oliver hear the story. We will listen to the story from both of you! How does this sound?" Betty added.

"Why me and why not the children? They can spend some time with their father," said Sofia, as a faint smile appeared on her face.

"First, this is because you two adults can pay much more attention and take lessons from the story, and second, the show is in the evening, the kids are most going to sleep and not listen to the story, anyway."

"Mom, do you have the air on? It is getting boiling in here," said Oliver.

"Oliver, do not make me laugh. But you adjusted the temperature when you arrived," said Betty as she laughed.

"What can I say? It is Mr. Anderson's call," said Sofia.

"Um, okay; we can go, but on a few conditions," replied Oliver.

"Let's hear them."

"First, you will not call me 'Mr. Anderson' from now on. Second, you will give me the honor to pick you up from your apartment; and the third is, we go out on dinner before going to the event. What do you say?"

"Okay Mr. Anderson—I mean, Oliver. I will text you my address," said Sofia.

Oliver beamed.

"I am glad that you both will spend some time together," said Betty.

On this, Sofia blushed.

"Well ... I would like to take a leave now. Have a good night. I say goodbye to the kids and then leave," said Sofia.

"We wanted you to stay," added Betty.

"I am sorry, but I need to go."

"Okay sure. Oliver, my dear!" said Betty.

"Yes, Mother?"

"Take her to her car, please, and bid her a warm goodbye," instructed Betty.

Sofia blushed again.

"Sure, Mother," said Oliver as he guided Sofia toward the main entrance.

Sofia hugged the kids and bid goodbye to Oliver. She got in her car, and off she went. Oliver kept waving until her car was out of the main gate. He began to see a potential life partner in Sofia.

Betty saw this by the window. She looked up to the sky and said, "Oh Elizabeth, as much as I wanted you to be here with my son and the kids, I am happy my son has started finding happiness in other people. You are also a mother, and you would understand why I pushed him toward Sofia. She is a good woman. I don't know how long I will be around to look after him. I couldn't leave him like this. You, my child, were irreplaceable."

"Talking to yourself, Grandma?" said Leslie from behind.

"No, my child. Come, let me tell you a story and put you to bed," replied Betty as she took Leslie into the bedroom.

7

Opportunities

The plane touched down in Baltimore, Maryland. Ariana got off her plane, looked for her luggage, and headed toward the exit. As she stepped outside the airport, she saw her sister Laura with her husband Henry and their cute daughter Lily. As Lily saw Ariana coming, she ran toward her. A big smile appeared on Ariana's face at the sight of Lily coming toward her. She got on her knees and hugged Lily tight. Once she was done hugging and kissing Lily, Ariana got off her knees and hugged Laura and Henry.

"Three years, my dear sister, three years; it felt like a lifetime without you," said Laura as she hugged and squeezed Ariana in her arms. Both sisters had tears in their eyes.

"Let's go home, ladies; you girls can have the rest of the day weeping and hugging each other," said Henry as he also hugged Ariana again and took Lily in his arms.

The reunited family walked toward the car. They got in and went home. As Ariana entered her sister's home, she felt like it was what she needed: family. She missed her family the most; she had not come to this place for three years, but it felt like she had never left. Ariana stood in the hallway, getting accustomed to the place.

Lily ran from one corner to another. "Ari is home, Ari is home," she shouted with great enthusiasm.

"Come inside, sweetie," said Laura as she held Ariana's hand.

"I was just looking at everything. They've all changed, but it feels like I know every corner of this house. It is so good to be back; this is what I needed," said Ariana as she pressed Laura's hand.

"You know you have come at just the right time. I am already two months pregnant, and my baby deserves to know Aunt Ariana loves her a lot," said Laura.

"Oh, my God! I am very excited about your baby," replied Ariana. Her face was beaming.

"Tell me about yourself, little sister. How have you been? How's your personal life? And how is your job treating you?" asked Laura as she took Ariana to the family room.

"Everything is fine, except that I have had some small challenges at work. You know, there are people who cannot stand the fact that you are also doing your job and getting appreciated for it. For them, they will always find an excuse to say something negative and make the other person feel bad. Professional jealousy is hindering my breathing space at work," said Ariana.

"I understand, my baby sister; those kinds of people are everywhere."

"But this is the negative influence that demotivates me, a lot!" replied Ariana.

"We can do one thing for you," said Laura. She was all charged on account to putting this idea forward.

"What?" asked Ariana. She too wanted an escape.

"Answer me this question: Would you like to move back to this area? That's how we are close. I cannot wait another three years to see you again. I know you want your privacy, but here's a place to stay until you can stabilize and live wherever you like."

"I ... I have never thought of this," said Ariana. She didn't have any words. This idea had never struck her mind.

"I already pitched this idea to Henry, and he likes it, so please Ariana, considers this once," impressed Laura.

"I will see. I need time to think about this," replied Ariana.

"Apart from being my younger sister, you are also my only sister and I would not like something to happen to you living at a distance. I would like you to be as close as before when we lived in Florida. I spend all day thinking about you, your health, and your happiness. I want you as close as possible," replied Laura.

"Okay, I will consider this," replied Ariana in a low voice.

"Yeah!" Laura clapped her hands.

"Let's go to the kitchen and cook something delicious; after three years, my beloved baby sister has come," added Laura as she chortled with happiness.

"Sure, sis ..." replied Ariana. They headed to the kitchen.

***

The sun came up the next morning, shining bright. The sunlight came in from the glass windows and struck Ariana's face. She woke up. Finding herself in a cozy bed in her sister's house and away from the chaotic workplace condition had put a beaming smile on her face. She got out of the bed, took a shower, and ran downstairs. She was keen to spend this day with Lily and Laura. It was nine in the morning, and Henry had already left for work. As she came downstairs, she could smell delicious pancakes. Lily was sitting on a chair next to the kitchen counter, and Laura was preparing of the breakfast.

"Good morning, beautiful ladies," said Ariana in a loud voice.

"Welcome Aunt Ari!" said Lily.

"Good morning, dear sister," replied Laura as she put the pancakes on the plates.

"This was the best sleep that I had in years," said Ariana.

"I am glad," replied Laura.

"I have prepared your favorite breakfast, banana pancakes with caramel syrup, take a seat," she added.

"Could this day be any better?" said Ariana as she took a seat and began to eat.

Once they were done with breakfast, Lily went outside to play with the dog and Ariana helped Laura with the dishes. The sisters talked about old times, their childhood, and their parents.

Ariana told Laura her tales of the hospital. She shared her fears and how she felt alone. How she needed her family to be around but she wasn't able to reach them in due time.

"Ah, my sister, come, give me a hug!" said Laura.

Ariana ran into her arms. They were hugging each other when Laura began to feel something in her belly.

"Ow!" screamed Laura.

"What is the matter sis? Are you all right?" asked Ariana in concern.

"I ... I ... ow! I feel pain in my belly!" With this, Laura tried on lean by the kitchen counter.

"Wait, hold on," said Ariana.

Though she had seen many patients in her life, this was her sister. She began to sweat in tension. "I ... let me bring you a chair."

Ariana got a chair from the other side of the counter and made her sit. "Hold on, sister, let me call the ambulance."

The ambulance arrived in five minutes. Laura was screaming. Lily had come inside by this time. She was holding Ariana's hand and weeping. Ariana had instructed her to maintain her calm, and the baby was doing so.

"Will Mommy be fine?" asked Lily as she wept.

"Yes, she will be. Don't you worry; trust me!" replied Ariana. She was also concerned for her sister.

The paramedical staff checked Laura's vitals, put her on the stretcher, and drove to the nearest hospital. Lily and Ariana also went with Laura in the ambulance.

The paramedic put an oxygen mask on Laura so she wouldn't have difficulty breathing. On the way, Ariana called Henry to update him about the situation. They soon reached the hospital. Henry came in twenty minutes later and found Ariana waiting outside the emergency room, holding Lily in her arms, who continued to weep.

"I dislike this; they're taking too much time," said Henry. His face was pale, as if someone had drained the blood out of it.

"Henry, don't you worry! We must stay patient. If something bad had happened, someone would have come out and given us information. This is just a normal procedure. Laura must do well. They may do some tests, so they are taking time," said Ariana. Here, she was a bigger and mature person. She had seen such situations before and knew how to handle them.

The door opened. Two doctors came out, headed toward Henry and Ariana. However, one of them heads forward without talking to them. Henry stops the second doctor.

"I am Henry."

"Hello Henry, I am Doctor Jolin. How may I help you?" asked the doctor.

"My, my, my wife had been taken inside. Laura. Her name is Laura. Is she fine?"

"Yes, she is out of danger. Take a seat. I was coming to talk to you," replied the doctor.

"No, we cannot take a seat. Tell me if I can see her," asked Henry impatiently.

"Listen, please keep your calm. Listen to me," replied the doctor.

"Henry, please. We are all nervous but like these, we must remain calm, Laura will be fine!" said Ariana as she assured him.

"Thank you so much, young lady," said the doctor.

The four of them sat, and the doctor continued, "We asked Mrs. Laura some questions and she told us she had moved some heavy things in the house before preparing breakfast. I want you to be very careful with her and pay attention to prevent her from doing it again. Also, we found that the baby had moved from its position. That she had moved some heavy objects had caused all of this. Right now, she needs rest; we will keep her under observation for a few hours and then let her go. We want to be sure the baby, and her, are not in danger. Questions?" asked the doctor.

"No," said Henry.

"Any prescriptions?" asked Ariana.

"It is unnecessary. Since she has already gone through the pain and it has also subsided, we don't find it important," replied the doctor.

"Thanks, Doctor, for the information," said Henry.

"You can see her, but only the two can enter at once to meet the patient. The other person may go inside once they come out," instructed the doctor.

"You two go first. I will follow," said Ariana.

"Well, I will leave you. Have a happy rest of the day."

"Thank you, Doctor for everything!" replied Henry. The doctor then walked away.

While Henry and Lily entered Laura's room, Ariana stayed in the waiting room until they came out. During this time, she heard two nurses talking about the need for more staff in the department. Ariana got up from her chair and went to where the nurses are talking.

"Greetings, I could not help but listen to the conversation you guys are having and it caught my attention. My name is Ariana. I live in Florida and I am a certified nurse."

"Nice to meet you, Ariana. I am Nurse Abby. How can we help you?" replied a nurse.

"I heard you both talking about how the department needs more staff. Where can I find more information about positions?"

Ariana began to consider the option of staying with her sister.

"You can go to the hospital website and there you will find more information about it. Our department needs more staff to fill some vacancies," replied the other nurse.

"The lady talking to the gentleman in the wheelchair is our supervisor, and she is a nice person. We can introduce you to her and you can ask more questions about it," added Abby.

"That will be great! If it's not too much trouble, I'd like to talk to her," said Ariana.

"Oh yes, and look, she is coming here," said Abby.

Soon, the supervisor came into proximity.

"My girls, is everything good?" asked the supervisor. Her composure was warm, and she sounded friendly.

"Yeah, everything is all right. We want to introduce you to Ariana; she is a certified nurse and lives in Florida. She wants information about positions," said Abby.

"Oh, hello, beautiful lady, I'm Dr. Jaqueline. First, girls, you two need to get back to work and check on two patients in rooms fifteen and twenty. Don't wait, run!" replied the supervisor.

"It was nice to meet you," said Abby.

"I hope to see you again soon," said the second nurse.

"Thank you very much for your help; see you," replied Ariana.

"We are very busy today, and this will continue because of the lack of staff. We have several positions in the department. If you are interested, you can go to the hospital website, see the positions and apply." She took out a card. "Here, this card, when you apply, call me to schedule an interview, what do you think?" asked Dr. Jaqueline with a broad smile.

"Perfect. Thank you very much for your time," replied Ariana.

"No problem, that's what we are here for: to help each other. Do you have family in the area? Since I believe you live in Florida," asked Dr. Jaqueline.

"Yes, my sister lives here with her husband and daughter. Their house is close to this hospital."

"That is good! It has been a pleasure to meet you Ariana. The hospital offers good benefits, including a retirement plan. Also, if you want to continue studying, they will cover twenty-five percent of the studies. Think about it. I assure you that you will feel like family. I hope to hear from you soon. I have to leave you. See you later," said Dr. Jaqueline.

"Thanks again," replied Ariana.

At the same moment Dr. Jaqueline left, Henry and Lily came outside to let Ariana met her sister.

"I was so tense. Why did you have to do this?" cried Ariana as she sat by Laura's side.

"See, so I need you to stay with me. Do you get me now?" said Laura as she smiled.

"Oh, sister. We will talk about this later, you get on your feet and then demand anything," said Ariana as she planted a kiss on Laura's forehead.

After three hours, Dr. Jaqueline was the one who completed the final procedures before Laura could go home.

"Remember that you need to rest. Lady, take care of yourself and the baby," Dr. Jaqueline said while pointing to her belly.

Laura nodded.

"You cannot be lifting heavy objects. Also, I would like to instruct all the members of the family please take care of her and make sure she raises nothing heavy. I need the signature of an allowed person," Dr. Jaqueline instructed Henry and Ariana.

Henry signed the documents. They were all set to leave. As they began to walk out of the room, Dr. Jaqueline stopped Ariana and said, "Ariana, I hope to see you soon."

All of them sat inside the car and the journey to home started. During the journey, Laura asked Ariana, "Where did you meet Dr. Jaqueline?"

"I met her outside. Two nurses introduced me to her. I wanted to talk to her for vacancies in the hospital and she told me that there are positions available here."

"This is the happiest thing to hear today; you just made my day. It's the best news I've been able to hear all day," Laura said as she rejoiced.

"So, Aunt, are you going to come and live with us?" asked Lily with full enthusiasm.

"It would be wonderful if you came to live with us," Henry added.

"No, not like that. If I move to this area, I can live with you for a while, but then I want to have my apartment to have privacy. I like it that way," added Ariana.

"I'm worried about Laura staying alone in the house," replied Henry.

"I've been thinking about it. I realize my sister does not rest. She always has something to do, and that worries me."

"Do not worry so much about me. I'll be fine. Regarding you, Ariana, it would be a joy to have my sister close to me," replied Laura.

"I will go into the hospital's website and consider the positions they have available," said Ariana as she patted Lily on her shoulder while she slept in the backseat by Ariana's side.

As they reached home, Henry opened the door for Laura to get out of the car; he helped her to take off the belt, and took her toward the door. Ariana carried a sleeping Lily in her arms.

It had been a rough day for all of them and they all needed rest.

***

Tuesday, life resumed for everyone except for Laura. She was being given special protocol by Henry, Ariana, and Lily.

They tried to do everything for her they could in the best of their capacities. Henry took a day off from work and helped Ariana with the house chores. They were both determined to let Laura rest at her best.

They couldn't keep track of time.

"God! It is already 2 p.m.!" blurted Lily as she received a call from one of her friends' mother on Laura's phone.

Laura answered the call while Ariana and Lily waited.

"I will let you know soon," replied Laura before she hung up.

"It was Mrs. Garrison, Esther's mother. She was inviting me and Lily for a show that will take place at Williams High School, where Esther's Brother goes," Laura told Lily and Ariana.

"Can we please go, Mom? Please, please," said Lily.

"She has already taken extra tickets for anyone who would like to join," added Laura.

"Then we need to go, Mom!" added Lily.

"Your father will work late this Friday. However, we can go if Ariana will drive," replied Laura.

"I will go if it makes my sister and Lily happy," replied Ariana.

Lily clapped upon hearing this.

"Yes, I can drive. You just have to guide me because I do not know the area. I am visiting this place after three years," added Ariana.

"No worries, sister! I will guide you, you just take out the dresses we will wear," replied Laura.

"I am already excited," replied Lily.

"Sure. I will call Esther's mother and tell her to keep aside three tickets for us," said Laura. "We can take them from her on Friday," added Lily.

"Okay my dear!" replied Ariana as she sat by Laura's side.

***

Thursday, it was a big day for Amanda and Jeffrey. Jeffrey would interview Amanda. 1,500 women were attending this show. Everyone was all set. This had to be one inspiring interview.

Apart from that, Jeffrey would interview the woman he loved. The camera began to roll and Jeffrey gave a thumb up! With this, the show started.

"Thank you, everyone for your presence today. It is a pleasure to have you all here. And thank you, Amanda, for taking out some time for this interview. It is an honor to have you here, a successful businesswoman who has been an example for many to follow. You are someone whom our young generation needs as their role models, be it girls or boys, you are a true inspiration of never giving up for everyone!" said Jeffrey.

"Thank you," replied Amanda.

"Audience, Amanda is a single mother, a fighter, and someone who has gone through difficult times and despite everything, has not given up, and has always been victorious," said Jeffrey as she turned toward the audience.

The audience applauded.

"The honor is mine. Having a journalist like you interviewing me is something that doesn't happen to everyone. Thank you so much!" replied Amanda.

"For those who do not know, Amada has a cosmetic company; her company is a supplier to retail stores throughout the USA and is still expanding. Tell us, how hard it was for you to start your own company?" added Jeffrey.

"At first it was difficult. Many now see a successful woman but do not know everything that I had to go through to reach this level.

When I started this business, I started selling cosmetics in my neighborhood; imagine visiting each house in the neighborhood and talking about the new products that had come out. The technology was not developed as it is now. I started this venture knocking doors. I depended on the direct connection with the customers."

"Let me tell you all a few things about this businesswoman's background: she came from a poor family; after she graduated from high school, her parents didn't have money to send her to a good college. Amanda, why don't you tell us how you broke through these barriers? Tell us about your childhood and how you started selling cosmetics," asked Jeffrey.

"When I was little, my parents told me I was talking too much and that I could work in marketing. There came the first idea. My mother was a housekeeper, and my father was a dishwasher in a local restaurant; they didn't earn much. The only time I could see my parents was at night when they got home. I was keen to turn the tables. The money that they were making was only for surviving—paying rent, bills, food, and saving for an emergency. They did not have money to send me to college. However, I was keen to make my identity. They told me that if I wanted to go to college, I needed to make money to pay for my college on my own."

"Ooh," said the audience.

"I didn't want to end up as my parents did, so I had to find a way. I started selling cosmetics in my neighborhood. After the first year, at nineteen, I made around $15,000 by selling cosmetics."

"Marvelous," said Jeffrey.

"That was a lot of money for a nineteen-year-old. Then, my father got sick and had to resign from his job, so his share of money disappeared from the house, and they added his medical expenses to the list. All their savings were used and then I had to help my mother to cover the expenses of the house, from paying rent, buying food, bills, and my father's medicines—I paid for everything. The money and hope for college was gone. I stopped selling my products and got myself a job. After a few years working as a housekeeper, I found that I didn't want that kind of life. I wasn't made for this. One day, I remembered when I started selling cosmetics again."

With this, Amanda paused. She had been speaking for too long. However, everyone, including Jeffrey, wanted to hear more of it.

"Then, what happened? I am impressed. Please continue," exclaimed Jeffrey.

"Then, at thirty …"

Amanda then paused.

"Are you okay?" asked Jeffrey.

Amanda began to think about the time when she was married and in love with the man that she thought would spend the rest of her life with.

"Mm …" Amanda was confused.

"You said that at thirty …" added Jeffrey. He wanted Amanda to let it all out. He wanted Amanda to get over her past. This could only do this if she would address it out loud. Thus, Jeffrey kept pushing her.

She continued, "At thirty, I started selling cosmetics and beauty supplies again. In the same way, knocking on my neighbors' doors and going everywhere. It was difficult because by that time I was married with kids. I had more responsibilities. Things took a steady pace. After six months, I began to see results. I started to organize myself. I had to be in different places to sell more products. I started having more clients as time passed. The demand began to skyrocket. Because of the high demand, I felt forced to look for a location, and from there, I began to develop the vision of the business that I wanted."

"This was a test of courage," added Jeffrey.

The audience continued to listen.

"Have you always received the support of people in your community?"

Amanda sighed. "What is a success story without struggles?"

"But hadn't you struggled for much already?" asked Jeffrey.

"I did, but this was another strike on my confidence. Many made fun of me and said my business would not last long. They told me they can guarantee my business would fail. Others said my children would be an impediment to the development of my business. They said childcare and career could never go hand in hand. Others who had already tried that business said people would tire of buying the products and I would end up in debt. All along, they had different scenarios in their mind projecting only one result: my failure. At some point, I too began to fear and thought about quitting. People always try to discourage you, but the difference between the successful and the unsuccessful is the decisive power and the will of never quitting. Successful people make the right decision at the right time. This is what I did. I didn't quit."

"It was worth it; all this effort, it was all worth it. You have raised your business from scratch. We are all proud of you," said Jeffrey.

"This interview has been insightful. My final message for you all is to not let people who have not achieved something in life be your counselor. Just do not listen to them; do not let their failures guide you or be your torchbearer. They are only good at killing dreams. Life is full of opportunities; do not let it go. This is just one life, so make the most of it."

Jeffrey addressed the audience as he turned toward them. "Let's give a round of applause to Amanda. She has done what most couldn't. She has broken the stereotypes, and she has made us proud. This is the end of the interview. We thank you, Amanda, and we thank everyone for being here today."

With this, the camera was turned off, and the interview ended.

A few people came on the stage to request an autograph from

Amanda and Jeffrey and the rest made their way out. Backstage, Amanda and Jeffrey were served drinks before they leave.

"It was a good interview, Amanda," said Jeffrey.

After sipping the drink, Amanda smiled and said, "Yes, however, you have one more interview left—at Williams High School."

"And you are coming along. We can have dinner before or after the event; it will be your call," said Jeffrey as he smiled back. He was beaming at this thought. She could see it.

"Before the show," replied Amanda.

"Then it is a deal," said Jeffrey as their glasses chimed together.

***

As per the plans, Peter, Jessica, and Coach Jengo Roger went to meet Ruth Soliman. She was the mother of Albert and Joel.

Mrs. Ruth was rather a humble and hardworking woman who tried her best to make ends meet.

"Hello, Mrs. Ruth. This is Peter and Jessica Johnson," said Coach Jengo.

"Greetings to all of you. Please come inside," said Mrs. Ruth.

She took them to the living room. "Would you all like something to eat?" asked Mrs. Ruth.

"No, thank you. I would like a glass of water," replied Peter.

"Me too," said Jessica.

"A glass of water for me," said Coach Jengo.

Mrs. Ruth brought water for all them, sat by Jessica's side, and said, "Coach Jengo told me you guys need to discuss something important with me, and it's about Albert and Joel. Please tell me about it. I'm concerned."

"Well," said Peter.

"Coach Jengo will guide you through the entire situation," added Jessica.

"Okay," Coach Jengo was about to begin. "See, the matter is, as you know, their son, Ariel, is the captain of the basketball team at Jefferson. The same team Albert and Joel are part of. The team couldn't make it through the semifinals."

"Yeah, so?" asked Mrs. Ruth.

"Albert and Joel have been bullying their son, Ariel. They even made public statements against him and have been threatening him since the match ended. Though I understand that we all wanted to make it to the finals, trying to lower someone's self-esteem isn't the right thing to do," said Jengo.

"I understand. This is unacceptable," added Mrs. Ruth.

"Let me show you the video where they have called Ariel out." With this, Coach Jengo showed her the video of her sons.

"Oh my God. I am very sorry, Mr. and Mrs. Johnson," said Mrs. Ruth.

"Our son is under immense pressure. They hurt him. He already wants to give up," said Jessica.

"Oh no! Your son is a gem; he was the star player. It was because of him that team even made it to the semifinals," said Mrs. Ruth.

"See, you also know the kid's potential. He shouldn't be brought down like this," said Coach Jengo.

"I am sorry. I should have been a better mother," said Mrs. Ruth as she looked down.

"Do not say that please. What the kid does outside remains unknown to the parents," said Jessica.

"No, it is my fault. I have to do two jobs to meet the expenses of the three of us. This is the reason that my kids have been neglected. They were innocent children like every other kid, but they have had a tough childhood. They always saw their father beating me and violating me. They have always seen violence in the house, and their father drove drunk and ended up killing a Dominican named Carlos. He is serving his jail sentence," said Mrs. Ruth as she wept.

"Oh God. I am very sorry to hear that," said Jessica.

Peter and Jengo looked at each other, but didn't say a word.

"No, I am sorry. This has made my sons bitter beyond measure. They are inflicting upon others what they had inflicted on them inside the house. They saw their father dating other women, even bringing them home and insulting me daily has taken its toll," added Mrs. Ruth as she continued to cry.

"Mrs. Ruth, please, be strong. It is only you who can look after these kids. We didn't mean to hurt you. We want the good for your kids and for our kid," said Peter.

"I am sorry that I began to cry like this," said Mrs. Ruth as she wiped away her tears.

"No, you do not have to be sorry. We all need to vent. Tell me if we can do anything for you," said Jessica.

"No, it is all right. I had to vent. I leave my home at six in the morning and return home just before midnight. I don't even get enough time with my kids. All of this has been frustrating for me," replied Mrs. Ruth.

"Knowing the reality behind their conduct, I can understand their situation. I will talk to my wife to help you get a job that can allow you more time with your kids. She has quite a few contacts and can help you with this. It is difficult to raise kids without their father. We are all at your service," added Jengo.

"If you need anything, if you ever feel enjoy talking, please keep my phone number. You can call me any time," said Jessica as she gave Mrs. Ruth her contact number.

"We shall take a leave now; we do not want to occupy all of your free time. Give lots of love to the kids and please don't scold them for what they did. Just talk to them about all of this. We don't want them to get hurt," said Peter.

"Yes, please, it is a sincere request," added Jessica.

With this, all three got up. Jessica hugged Mrs. Ruth and Peter and Jengo shook hands with her. All of them left the house. The Johnson family bid goodbye to Jengo and also thanked him for coming along.

Coach Jengo went home and so did Peter and Jessica. They had to prepare for the show at Williams High School on the next day. Ariel was waiting for them at home.

***

Friday soon approached. Laura's condition was much better. They were all set for the show. Laura was dressed in pink and Lily and Ariana were twinning for the day, wearing yellows.

"Both the girls look beautiful!" exclaimed Laura.

"Thank you, beautiful sister!" replied Ariana.

It was a fine Friday, and the weather was refreshing. The three ladies sat in the car and Ariana drove.

On the way, Laura guided her with the route. Lily sat in the backseat, playing games on the phone.

"The show will start in the evening. I think we have left the house too early," said Laura.

"Then I think we can eat something. What do you say?" asked Ariana.

"Ice cream!" screamed Lily from the backseat.

On this, Ariana and Laura began to laugh. "Okay young lady!" said Laura.

"There is one ice cream parlor that serves the best ice creams in town. I will tell you where to stop," instructed Laura.

"Okay," replied Ariana.

Ariana stopped the car by an ice cream parlor as instructed by Laura. "Which one do you guys want?" asked Laura as she stepped out of the car.

"Mango," said Lily.

"Blue moon," replied Ariana.

"Okay, I will bring both," said Laura as she closed the car door and went inside the ice cream parlor.

Ariana and Lily waited in the car as Laura came back. She opened the door and said, "Girls, let's walk to the park. It's nearby. We have some time before the show."

"Yes, sure," said Ariana as she and Lily got out of the car.

As they entered the park, Lily asked Laura if she could go to play.

"But who are you going to play with?" asked Ariana.

"Let's see. I will find someone to play with me," replied Lily.

"But don't go too far," said Laura.

Lily found two kids to play with. Laura and Ariana took a spot on the grass as they watched Lily playing.

"So tell me?" started Laura.

"Tell you what?" asked Ariana.

"About your personal life. Do you still have a crush? Or a guy you love?"

"That was a tale of last year. From that point onward, I've been concentrating on my work. I don't have time to fancy a guy. There are many who are looking for girls to play with their feelings. I cannot waste my time like that," replied Ariana.

"But you also are struggling with work. Seeing someone you like would bring positivity in your life, but I get your point," replied Laura.

"No, please give me any other suggestions," replied Ariana.

"Change your workplace. Maybe move here?"

"I'm afraid to start a new job. I already have given about three years to the place where I work, and although things are not going well, at least I know my responsibilities and the staff. I am familiar with that place. I am afraid—too afraid—to start over. I do not know how the work will be and how they will treat me, which is why I have stayed where I am."

"But you are a hardworking girl; your work will speak for itself, no matter where you are," replied Laura.

"I get your point. The best times for me are the times to mark my card and go home, payday, and when the time of departure comes for my days off. Do these count as job satisfaction?" asked Ariana.

"Are you considering the jobs at Maryland Hospital?" asked Laura.

"No, not yet," replied Ariana.

"Mom, Aunt, what are you guys talking about?" Lily had come back.

"We had a wholesome conversation, sweetie. Come on, it's about time. We should get going. It's five o'clock already," said Laura.

"Sure. I cannot wait to hear the story! I am super excited," replied Lily.

"We will also grab something else to eat before the show, so let's get rolling," said Laura as they went back to the car.

Laura, Ariana, and Lily returned to the car and were all set for the show.

Laura turned the music player on. It was a good journey. "I want a hamburger!" said Lily.

"Okay, my dear, as soon as I see a place for fast food, we will stop and buy ourselves the dinner," said Laura as she looked at Lily in the backseat.

"Yeah, just tell me where to stop," said Ariana as she continued to drive.

8

The "Friday" Affair

It was the time when everyone's life was deemed to change. The press was covering the show at Williams High School.

Jeffrey, the chief reporter covering this event, was the center of attention. Amanda was sitting in the front row and had a seat reserved for Jeffrey by her side.

Laura, Lily, and Ariana had a seat in the second row. Emma and her sons were in the back while Ariel and his parents were in the third row.

The event hall filled in a few minutes' time.

Jacob, Aidan, Oliver, and Sofia were among the last ones to enter the hall.

Jeffrey came onto the stage, introduced himself, and said, "Respected audience, this is to inform you that the show will start soon. Uncle P is here among us. Give him a big round of applause."

With this, the audience gave a round of applause as an old man stepped onto the stage. This man was sixty-five years old.

He introduced himself as Uncle P and told the audience he was the new Spanish teacher at Williams High School.

"Hello everyone. I am Uncle P. Maybe you would like to know what the 'P' stands for. The 'P' stands for my last name: Phipps. I have moved from the Dominican Republic to the United States. Today, I will share a story about a village I have titled 'Challenging the Unknown.' You all must think I haven't told you my real name. I find it needless because a person is known by the deeds he does and not his name. Today, I wish to tell you guys why it's important to get out of our comfort zone and explore the unknown."

The audience listened to him with their undivided attention. Uncle P then sat on a chair in the center of the stage. A school assistant had turned all the lights of the event hall off and turned the one above Uncle P on to highlight him among the masses. From here, he began with the story.

"Once upon a time, there was a village in the middle of a forest where his inhabitants lived peacefully. They would sustain themselves for most of the necessities of life, and they would trade amongst themselves to get whatever they lacked. Money was not used as a mode of exchange; people bartered. One rule that the villagers had was that no one would go outside the boundary of the village because of threats of the animals outside the village. This was because when people would go out of the village, they would often become the prey of the animals surrounding the village."

He immersed everyone in the story.

Uncle P narrated:

The village was so quiet during the nighttime that the inhabitants could hear the wolves and the tigers. In the village lived a witty boy named Joseph. He would always have questions regarding the conventions followed by the villagers. Each night, whenever he looked up to the starry sky, he had questions for his grandfather. One night, when Joseph was almost ready to go to sleep on his hammock, he asked his grandfather Herald, "Grandpa, why are we surrounded by many trees and ferocious animals?"

To this, his grandfather replied, "This is because we live in the middle of a forest."

"No, we live in a village," corrected Joseph.

"And our forefather built this village in the middle of a forest my child."

"How come?" Joseph was surprised.

"Many years ago, our forefathers traveled in packs to search for food and shelter. They traveled miles with their families. Many people died in the course. They came across this place, in the middle of the forest, and found all they need for the living. However, it wasn't safe to live here. Thus, they created a zero-shaped boundary outside the village. The elders prohibit the villagers from crossing it."

"Why?" the witty Joseph asked again.

"Our ancestors believe that anyone who crosses the territory will either fall prey to the wild animals outside or end up being lost in the forest. They fear what is out there; the villagers title anyone who thinks about crossing the boundary as crazy."

"Oh, but have anyone tried to do so?" asked Joseph.

"A few people did, but they couldn't cross the forest, most of them returned and were humiliated. The villagers mocked them for the rest of their lives."

"And what happened to those who never returned?"

"The elders told us they died on their way out," said Herald.

"If someone has the right strategy to execute a mission, they can emerge," said Joseph while he still looked at the starry sky.

"You are right, my grandson."

"You know what? When I'm the age the village requires, I'll leave this place and cross the forest. Grandpa, do you think I could do it?" said Joseph.

"Yes, I know you can do it my grandson. Anyone with the right strategy can do it. You know what went wrong with the people who tried this before and did not a success?"

"No, what?"

"They tried to leave the village but returned because in their mind they had already accepted that the unknown of forest would defeat them. The thing one should be concerned is when taking a step different from what the crowd says is, if they were to do it, why will they be able to achieve it? What difference will it make? And how will it be beneficial for you and for everyone around you? You need to understand that spiritual enlightenment needs to be achieved rather than physical success; if it makes you a better human being, if it makes you wise and if it makes you knowledgeable, it is worth fighting for," instructed Herald.

"Wow."

"Today I will share something that I told no one, not even to your father."

"What is that, Grandpa?" asked Joseph.

"Look, today I am an old man. There were many things that I wanted to do in my life, but I didn't. I didn't make many decisions in my life because I feared failure."

"What failure, Grandpa?"

"I don't know; those failures are unknown to me because I never took those chances. I also feared the unknown and deprived myself of something that may have resulted in success."

"Ooh!"

"Look, Joseph, there will be many people who will only be there to discourage you; however, a few will be there to encourage you. Do not forget that life will be full of people who keep on discouraging you, who will tell you that you shouldn't take a chance, and who will tell you it is not worth it. However, it is upon you to understand that some people and some opportunities are worth your time, in the path of your life, you will find many people who will need your help and this may slow down your progress, but it doesn't mean that you do not help them. A successful human being helps himself and the surrounding others, only your character, decisions and discipline will separate you from the masses."

"But Grandpa, what if I run out of time when I help others?"

"You won't. Everything happens for a reason; we just need to embrace the opportunity when it comes. That's it; people may try to use you for their personal benefit just because you are showing kindness, but know that your kindness will always be reciprocated either by people or by God."

"Okay, Grandpa. I am feeling sleepy now. You always have gems to spill; your words motivate me a lot. Good night."

"Good night, my child," said Herald as he caressed his grandchild's cheeks, who went into a deep sleep.

Grandpa began to look at the starry sky with the same shine in his eyes as Joseph.

He then said in a low voice, "Oh, Joseph, I know that you can; you only have to be convinced of what you will undertake and see the end before it arrives. Understand that we need to take the right decision at the right time. Time flies and we grow old. What we can do today may not be tomorrow. If we do not understand this, this life will become harder to live and hardest to continue; nothing is worse than living with regrets."

The next day, Joseph woke up, and after having breakfast, he went to play with his friends. He saw the big gate of the village that served as an outlet to the forest. He said to himself, "One day, I will cross this gate and challenge the unknown."

When Joseph said those words, he did not realize a few feet away, a young man heard him say all of this. He came up to him and began to laugh in his face.

"Who are you? Why are you laughing?" asked Joseph.

"In the village, I am better known as Scarface. You know why, kid? Because, on one occasion, I tried to leave the village in a desperate search for a way out. After a few days, I met a hungry lion. In my escape from death, I did not pay attention and was hit with a branch of a tree that gave me this wound and upon healing, it became a scar," said Scarface as he pointed toward an ugly scar on his face. "Do not even try, or get ready to be treated worse than me. Hahaha! A kid like you? Hahaha!"

Joseph looked at him with surprise and confusion. He watched Scarface laugh. As he was being called by his friends to return to the game, he left without saying a word. However, what Scarface said echoed in his mind.

One of his friends asked, "Joseph, what took you so long?"

Joseph didn't answer.

He shook him and asked again, "Joseph, what are you thinking?"

The shake brought Joseph back to reality. "Umm! Just thinking about making my way out of the village grounds."

On this, the kid began to make fun of him. He chortled and gathered every other kid in the ground and told them that Joseph was cracking the best joke.

"Will you go through the forest?" asked one kid.

"Then good luck," said another.

All the kids laughed.

Joseph's mind was boggled. Other kids younger than him made an issue about this topic. He had never known its sensitivity before, but now he did.

"If you return, the villagers will humiliate you just like the others. Or maybe you will fall prey to the animals in the forest. Good luck fighting tigers and wolves!" added another kid.

"Just like his father, he will never show up again to the village. He too had fallen prey to the unknown species in the forest. Ha! You will have an ending just like your father."

On this, Joseph jumped on the kid and began to beat him.

Herald, who was passing by with another man who was this kid's father, intervened. Upon being told about the situation, Herald asked Joseph to apologize.

"I will not," said Joseph.

Grandpa shouted to him and ordered him to apologize. On this, Joseph apologized, but as people were eyeing him with disgust, he ran away from the crowd. He ran toward the marketplace, took shelter under a tree, and began to cry. His grandpa, trying to follow him, came after him panting.

He held him with his hands as he panted. Since Grandpa was old, he needed some time to settle.

Joseph began to wail. "Why me, Grandpa? Why does it have to be me? Why my dad? Why not someone else's?"

On this, Grandpa hugged him and Joseph continued to cry on his shoulder. He said, "Joseph, right now I know you feel hurt by what that kid said about your father. But you must understand my grandson that does not give you the right to hit other people. People will say a lot of things and you will have to endure them with whatever effect they have on you."

Joseph began to cry loud.

"My child, no. Don't," said Herald as he wiped Joseph's tears.

He continued, "Do you want to know something, Joseph? In the path of your life, you will find people who make fun of you to make you feel bad. Many times, they will act with kindness, but then they will make fun of you to your face. I only ask you to stay away from these people. Many only want to provoke discouragement and sadness in you."

"But why do they have to be so mean?"

"Because they know you have the courage to break through all kinds of barriers while they will fear the unknown all their lives. They will not move, they will not grow, and they will not excel. Thus, they also don't want you to move, grow, or excel."

"I know, Grandpa. I was furious. Grandfather, why did my father leave and has not returned? Had he fallen prey of the unknown? It has been five years, and I have turned thirteen years old. If he is alive, do you think he remembers me?"

Many people in the audience were crying, including Sofia, Laura, and Amanda.

Oliver held Sofia's hand and said, "Thank God this is just a story, or you would have cried a river." Sofia wiped her tears just to give space to more tears on fall on her cheeks.

Uncle P continued:

On this, Grandpa said, "Oh, my son! We do not know what happened to him, but I'm sure he remembers you. I am sure that he is alive. He has a courageous heart as you do and he cannot fall prey to the unknown. You were his happiness and the driving force that kept him going through all kinds of hurdles so you would have a better future. He wanted to grant you the opportunity to dream and have a better future than the rest of the people who live in this village. After your mother became ill and died, your father was not the same person; he couldn't live life the way he used to; his face saddened, his behavior and character changed. He was not the same happy and helpful person who lived in the village. I also miss your mother. She was my delight and the light of my eyes: always understanding and affectionate. But we must accept reality. Your mother is not with us, and your father left the comfort and the village looking for a better future for you. I hope one day you will meet him again. Even if I am not alive, he will be there to take care of you."

Joseph wiped his tears and said, "No Grandpa; you cannot talk about leaving me. My father left, my mother left, and now you? I will never be the same if you leave. So, you do not have a choice to leave me."

A tear fell from Herald's eye.

Joseph continued, "Grandpa, if my father does not come, I'll go look for him. I will find him at any cost. This is not where I want to spend the rest of my life. I was not born to live the average life. Being born to Macy and Hudson, I cannot be average. Everyone in the village is satisfied and afraid to leave it. This strange satisfaction has taken over their ability to think out of the village. They do not want progression in their lives. I do not want to stay here. Someday I will go out, prepare, and return to offer a better future to the village—not for just myself, but for everyone. The great challenge is to get out of here and go through the forest of the unknown. It is the unknown that people fear; most of them do not even know what is in the forest. Tiger and wolves are the names they have learned from our ancestors, but they know nothing on their own. I know one day I will leave and will come back to make this place a city of dreams where dreaming costs nothing and preparing for a better future is an opportunity for anyone who seek it."

"Joseph, my dear. I feel honored that my grandchild has so much to give to the surrounding people, and he thinks about everyone. People say that Dream City is a place with opportunities for everyone. I think if people join hands, they can turn any land into a city of dreams."

With this, Joseph hugged his grandpa and said, "I cannot thank God enough for blessing me with a grandfather like you."

From that day onward, Joseph planned to leave the village for a better life. He had a different charm in his eyes. His face lit up whenever he saw the gates of the village. His grandpa set him to learn the ways of survival: how to cut wood, light a fire, cook, wash, and clean. He began to equip himself for self-sustainment.

Between the ages of thirteen and seventeen, he volunteered to learn how to design and build houses. Throughout the village, they knew Joseph for his experience in designing and building houses. His creativity set him apart from the other designers and builders in the village. When he turned eighteen, he began to supervise construction himself. Though the villagers liked him a lot, he was determined to leave this place for good.

He would often tell Herald, "Grandpa, once I find a better living outside the bounds of this village, I will come back to improve the village. If they don't want improvement; then, I will take you with me."

Herald would always reply, "You and your father can live outside this village, but you will have to leave me here."

Joseph would always ask him why and he would never reply. However, Joseph got the answer one day. Joseph was on the construction site when a boy came running and told him, "Herald wants to see you as soon as possible."

Joseph ran to his house. He saw Herald on the hammock in a deteriorated condition, moving and struggling to breathe.

"Grandpa, what has happened to you? I will get a doctor," cried Joseph as he sat by Herald's side.

Herald stopped him by holding his hand and said, "This would not help my son. For me, my time has come. I want you to be strong. I want you to pursue your dream. This is my last wish. Bury me here, in this village, by Macy's side. I have yet to meet my daughter-in-law and thank her for the gift she gave me—that is you. Leave me behind and go forth in the search for what's better."

Joseph began to cry.

"No, Grandpa; come with me."

"I told you that you'd have to leave me behind. This disease will not heal. It will take me. My eighteen years with you were the best years of my life. You can make wise decisions on your own. I do not want you to stop yourself from achieving your dreams because of my illness. During the last few years, you have learned to make your own designs and build houses."

"But what if I need your guidance?" cried Joseph.

"The ones who love us never leave us, my dear. I will stay alive in your heart forever. I will look upon you throughout your life with Macy. I hope you are united with your father someday."

Herald began to cough. Joseph got up to pour him water, but he held Joseph's hand tight. He didn't let him get up.

He continued, "I need you to promise me you will pursue your dream and go out of this village for good, and you have to come back and visit my grave. Come back and tell me the tale of your victory, and you have to come back for your people and you have to help anyone who needs to be helped."

Joseph began to wail. He didn't say a word.

Herald held his hand tight and said in a loud voice, "Promise me!"

Joseph replied, "I promise, Grandpa. I will do everything as you have instructed me to do, but don't leave me now, please."

With this, his grandpa smiled and began to look up the sky. His eyes didn't move. They froze.

His grip on Joseph's hand stayed, but his hand didn't have the same warmth. Joseph cried, "Grandpa!" but Herald didn't move. Joseph freed his hand of Herald's grip and closed Herald's eyes.

Now he has reunited with his daughter-in-law in heaven.

Until this point, even Ariana and Oliver had cried. Uncle P smiled as he saw a lot of eyes weeping in the faint light. He continued.

Two months later, Joseph was ready to leave the village. Many asked him to stay while others made fun of him.

People told him that Dream City was a myth, and he would lose everything he has for something that doesn't exist. Joseph knew what he was going in search for. He knew those who cannot take the risk are those who say the city doesn't exist. For people of wisdom, the city is beyond the fears of unknown. One has to gather the courage to face the unknown. Only difficult roads lead to beautiful destinations; the term unknown creates an unexplainable fear.

Anyone who overcomes the fear of the unknown becomes entitled to success, but overcoming the fear of the unknown is not everyone's cup of tea.

The day when Joseph planned to leave arrived. He only equipped himself with a survival pack. He believed if he would need anything extraordinary, God would send it his way. Everyone comes to bid him goodbye. Most people still tried to stop him. Joseph told them he has decided and his decision is unchangeable.

Seeing this, Scarface, from a distance, said, "Poor boy; he is digging his own grave."

The village gates were opened and Joseph set out. He was keen to look for Dream City. Joseph looked back. The villagers had stay put in their places. They must have thought Joseph will change his mind and come back. Joseph remembered what his grandpa had said, then he remembered his mother Macy, and then his father. A tear fell down his face.

He took a deep breath and said to himself, "No Joseph, don't look back; this is your chance to make a difference."

With this, Joseph moved forward. The villagers who saw him moving instructed the guards to close the gates.

They began to talk amongst themselves that Joseph, like his father, will fall prey to the unknown and will never return.

"Ah! Someone has cursed this family in the past," said a villager.

"Therefore, either they died in their beds or gave themselves in the unknown's way," said another.

Joseph kept on walking with his satchel. Fifteen minutes later, he stopped at a point with two diversions.

"No matter how long and how difficult the road is, I'm willing to do what it takes. I promised my grandpa and I shall live up to it. I don't care what the unknown is. For if it is worth being concerned about, I will decide once I encounter it. From now on, I will not be the ordinary Joseph who was in the village, I will be the Joseph who will make a difference for himself and for the surrounding others. The Joseph, who will always be willing to help others, henceforth my mentality must be different, no matter if I have to be alone on this path, I will follow it anyhow," said Joseph.

After that, he took the diversion on the right, which would take him to Dream City. As he passed through bushy trees on his way, it became silent. There was not even an animal nearby. He could hear the wind and his steps as he stepped on the grass, and then on broken wood, and then on dried leaves. Suddenly, he heard something in the bushes, as if someone was approaching. Joseph took out a spear in his hand and assumed a position of defense by the tree to protect himself.

"Who is there?" shouted Joseph.

That thing making the noise became still. Joseph picked up a rock and threw it at the bushes.

On that, a young man of the same age as Joseph came out running. "You could have hit me on the head!" said the young man.

"Who are you?" asked Joseph.

"For those who know me, I am Jonathan; for those who don't, I am a dreamer," replied the young man.

Joseph chuckled.

"Like you," he added.

"Like me? Have we ever met somewhere before? I find you familiar," said Joseph in surprise.

"Maybe, because we have a lot in common."

"Yeah, maybe. Can the two of us walk together without agreeing to meet?" asked Joseph.

"I agree," replied Jonathan.

"Look here! Have you been following me?" asked Joseph.

"No, no."

"Where do you come from, and where are you going?" asked Joseph.

"I'm going to the spectacular Dream City!" said Jonathan. His eyes began to shine and his face beamed.

"And where are you from?"

"Oh boy! You ask a lot of questions. You will get to know about it with time, but for now, we must advance before it gets dark," said Jonathan.

"We? I don't know where you have come from and, I don't know anything about you either. I need to be cautious. I need to make it to Dream City anyhow, and I cannot compromise anything in its way. If you become a hurdle, I shall go on my own and not let you accompany me," replied Joseph.

"Do not forget that could be many people on the way who may need our help and we cannot deny them. We can choose not to, but we shouldn't. I will not change myself because of the circumstances. I have one thing clear in my mind. I need to get to Dream City, but I will not forget about the people who may require my help on the way," said Jonathan.

Joseph eyed him with a perplexed expression.

"Be flexible for the path, but always be rigid for the goal," Jonathan added.

"Since when are you my advisor?" asked Joseph.

Jonathan smiled.

"I see you haven't asked me about my whereabouts," said Joseph.

"By the time on the road, I shall know," said Jonathan with a smile. Then they started walking.

After hours of walking, both the lads were thirsty, and it was already dark. They saw another village's boundary in the view of their naked eyes. They ran toward it. The guards at the gate stopped them and asked them who they were. Joseph was about to tell them but was stopped by Jonathan.

Joseph looked at the guards as Jonathan began to speak.

"We are residents from another village, and we are just passing by; we have been walking almost all day in this forest and we are tired and thirsty; we need a place to spend the night. Can we stay here before we are all set to start our journey again?"

The guards granted them a place to rest in the village.

They entered the village hungry and tired, and then they sat under a tree. Joseph had some loaves of bread that he carried with him from his village, while Jonathan had nothing.

A villager seeing the two young men noticed they were foreigners and had no place to sleep. He offered them food and accommodations. Others knew the residents of this village for their hospitality. They took it as an honor to help others. Joseph, who had never been out of his village before, knew nothing about them.

The next day, when Joseph and Jonathan were ready to leave, a strong storm hit the village. Call it fate or an act of God; neither of them could leave.

The storm devastated almost the entire village. Many houses were destroyed and people became homeless. Among the destroyed houses was the house of the elder of the village. His name was Imari. He was a fair and loyal individual and thus had assumed the position of the elder. Imari had two nephews whom he had raised. One of them was Deroc.

Deroc was an arrogant individual who never cared about others. He was always full of himself. He despised helping others. He would only talk to people or look into their matters if they were of any benefit to him.

Deroc had a brother named Hathath.

Hathath was one indecisive individual who would worry about the consequences before he took any steps. After the storm subsided, Joseph and Jonathan planned to continue the trip in the normal course, but seeing the situation of the village, Jonathan said, "Why don't we stay a few days helping to rebuild those houses? These people need our help."

"No, we cannot. If we halted like this everywhere, we will die before we make it to Dream City," replied Joseph.

Joseph continued walking. He wanted to get to the Dream City anyhow. He had left his settled life behind in search of a better future and stopping wasn't a part of the plan. After walking a few feet away, Joseph thought he could not let the people suffer like this; he will have to return and help them.

There was a lot of work to do and it took weeks and months to rebuild the houses. Joseph and Jonathan both helped in rebuilding the houses. During this time, they got to talk to Deroc and Hathath. Joseph disclosed in front of them he and Jonathan were going to Dream City.

After they had brought the construction to its maturity stage, Joseph and Jonathan left the village. Deroc and Hathath also left. They also wanted to find Dream City.

Imari didn't stop his nephews from going with them. He was a firm believer that one should be given a free will to decide between the right and wrong path.

Now, there were four people walking toward to Dream City. All four stayed alert the entire way because of the fear of the unknown. Hathath, who was the youngest among them, stayed in the middle because he feared an attack by the unknown the most.

After walking half a day, they stopped in front of a valley called the Valley of the Shadow of Death. Villagers said it that whoever tries to cross gets lost and dies with either hunger, dehydration, or falls prey to the wild species.

The four of them camped by the Valley and proceeded the following morning.

"If I had known, I would have stayed in the village. I felt safer there than in front of this valley. Look, we are exposed here. No one would know if we die. The howling of the wolves scares me and if something were to happen, we have no one to ask for help," said Hathath.

"Please do not start. You must understand life is full of challenges and we will always have to face it and our fears," instructed Deroc.

The wolves howled more as the night became darker. The wild beasts must have come out in the night. They could hear things moving in the pitch dark. After a while, they lit a bonfire and ate the food they had with them. Because they were so tired, all of them fell asleep in no time.

In the morning, they prepared their belongings and began to walk forward. They walked fifty feet, only to understand that the road had ended.

"Oh God, what should we do now?" cried Joseph.

"Maybe this is the end of our journey," said Hathath.

"Wait, let me see," said Jonathan.

He moved forward to see they had approached the edge of a cliff and then was no way forward. The forest was more than a thousand feet below and they couldn't jump. They had to go to enter the Valley of the Shadow of Death to move forward.

"Wait! I can see a path ahead that will take us to the Valley of the Shadow of Death," said Jonathan.

"Then come on. What can stop us? Let's continue our journey," said Deroc.

This time, they put Hathath forward. He had to lead the path down the cliff. Hathath, a young man with a fragile heart and the decisive power of a robot, halted on the way where the path became narrower. He refused to move. In that instant, the rocks from the top of the cliff began to fall; all four could hear them falling near of them. A landslide was coming their way.

Deroc shouted, "Hathath! If we all die, then it will be because of you … you have blocked the way. If you yourself want to live, then move! I said, move!"

"I can't," replied Hathath.

"You guys need to keep move. I cannot hold my feet on this ground for a long. Joseph, please try to pass over Hathath. I will fall from the cliff if you guys don't keep moving," shouted Jonathan.

Hathath was so afraid that he couldn't move. He had lost the hope to live. "This is our end, guys. I am sorry that I have risked all of your lives. If it wasn't for me, you guys would have made it." With this, Hathath left himself free to fall off the cliff.

At the right time, Joseph took hold of Hathath's hand and saved him. He took hold of a rock and asked Jonathan to move past the three of them. Now, Jonathan was leading. Deroc followed Jonathan and then came Hathath as Joseph held his hand and followed him.

"Do not look down. Look forward if you want us to get off this cliff. I will not let you die like this, trust me," said Joseph.

Hathath nodded and then began to move forward.

"You are doing very well," said Deroc.

"Do not stop. Just keep moving, Hathath," he added.

Jonathan, who was now leading, shouted, "We will have to move forward faster. I can see the rocks moving. I don't know if they will come down or not."

The four of them sped until they reached the end of the cliff below. After they came down the cliff, they fell to the ground and began to pant.

"Oh Lord! That was so close," shouted Jonathan.

"It is the survival of the fittest," added Joseph.

The four of them brushed themselves as they got up.

This was the point where people were annihilated or if they were lucky enough to make their way down the cliff alive, they would return to their village. They had no other option but to enter the Valley of the Shadow of Death. The residents of all the neighboring villages said anyone who entered that valley did not get out alive. No one has lived to tell the tale of the unknown, according to inhabitants of the neighboring villages. It was a valley where people died with their dreams trying to reach Dream City. Most people believed that since no one had made it past this place, it meant this valley is the ultimate end and no such a place as Dream City existed. The valley was a desert. They could see no one in proximity; only dried tree trunks.

As they moved forward, the boys could hear screeches and howls. However, no animal could be seen. Soon, they entered the valley. It was like the beginning of a new danger. The valley was misty. No one could make out anything past five feet, and the temperature was much warmer than the rest of the forest.

"This place is unexplainable," said Hathath. "I cannot believe there is a desert inside this forest. How come no one ever got to know about it?"

"If you live to tell the tale, then you will," chortled Deroc.

"This is because none ever tried to make it to Dream City," added Jonathan.

Soon, they began to hear something moving. "There is something around us, guys," said Joseph.

"Stay alert," cried Jonathan.

The mist in front of them dissipated, and they saw an animal wandering from one mark to another. "He is searching for food!" cried Joseph.

"What do we do now?" said Hathath. He was the first person to panic.

"Oh! Hathath; get ready," instructed Deroc.

The beast was huge and had long teeth. They took out their spears for defense and then tried to make out where the beast had entered. The animal had come from a bridge that looked small from a distance. All four of them began to run toward the bridge.

"Guys, put in all your effort. We need to run fast," said Jonathan.

"Grab large branches of trees after we cross the bridge, and together, we will face this beast. Let's defeat him once and for all," said Joseph in a loud voice.

Everyone acted as per the instructions apart from Hathath. He stayed hidden behind Deroc since fear had taken hold of him. Hathath, who stayed at a side, saw the fight as Joseph, Jonathan, and Deroc made the beast run away from them. All three of them were panting while Hathath clapped for them. They didn't like this; though they defeated the beast without Hathath, they still had to look out for him because he couldn't do that for himself.

After resting for a few minutes, they moved forward. They came across another village where they stayed for a while. This village was different from the one where the brothers lived. The people here were conformists, and anyone who would deviate from the told path was mocked and disliked from the rest.

The Dream City was like a mere illusion for them and they thought people who aren't able to live per the societal norms embark on the journey to find such a city and anyone who thinks about it only brings shame to his family.

These people feared changes. These people thought deviating from the set path would only lead to doom. Anyone who would speak of a new idea was feared by the villagers and would be kept in gates by the village custodians. They would deny such people public interaction. The villagers would conform to the norms so much that they lived in misery rather than taking steps against the misery to put an end to it.

They just had one answer to everything: "We cannot fight what has been a tradition."

The boys stayed there and collected the necessities for the journey was to be resumed.

"Deroc, why don't we stay here?" asked Hathath.

"What makes you say that?" asked Deroc.

"How do we know Dream City exists? If I had known that we

would face so many obstacles, I would have stayed in our village. I am afraid to continue, if we do not know what we will find on the road, and if the Dream City is a real city or a fantasy. I cannot risk my life more. I have had my share of adventure; this is all too much to endure."

"Then why did you come in the first place?" shouted Deroc.

"How could I stay in our village without you? I was too afraid to stay there without you," said Hathath.

"Do not say that, because that will not happen. I didn't leave that place to go back. Why do you want to stay? I don't get your reasons to have double opinions. These guys saved you, and now, you want to take the easy path and abandon them? For once, don't even think about them. Think about the life you can have in Dream City. God did not create us to live this kind of life. Fear will lead you nowhere! After we have traveled so far and crossed dangers, you still have the desire to go back? No, even if I have to continue alone, I will not go back," shouted Deroc.

"Dare you to leave me in this village?" Hathath had the same fear in his voice.

"I'm not leaving you; you want to stay here because you're afraid of what may happen later. See, most people said no one could cross the Valley of the Shadow of Death, but look at yourself; you are aiming through. You haven't fallen prey to the unknown. If you do not decide what you will do, look at how I disappear in the woods and toward Dream City. This is your time to decide," said Deroc.

"I cannot live here without you. I am terrified. I fear I cannot fake wanting to pursue this journey anymore. As much as I would want you to stay here, I think I should let the three of you go. I haven't been more than a hurdle for the three of you. You guys are brave; you can make it," said Hathath.

"You will regret it later. You will miss me, and maybe you can never see me again. You think you could have made it with us, but you didn't choose to. This will eat you from inside for the rest of your life," said Deroc.

Hathath said nothing.

Joseph and Jonathan were listening to every bit of their conversation. Finally, Joseph intervened. "If we already started this journey together, we will end together."

"Why to give up now that we are close?" added Jonathan.

No matter how hard they tried to persuade Hathath, he had decided this time. He stayed put. Between tears and hugs, Deroc said goodbye to his brother. "I'll come back for you," said Deroc as he departed from his brother.

"Do not worry. I'll be fine; you guys make it to Dream City, and live the life you want to. I love you," answered Hathath.

Joseph asked him if he was sure, as he continuously repeated the same thing about the reality of the villagers. He said, "The residents of this village have no aspirations for improvement and a better future, and they are accustomed to a style of conformism."

"I know, but I am not brave like you three. I cannot come, you guys shouldn't waste your time in persuading me, go," said Hathath.

After these words, Joseph, Jonathan, and Deroc continued the journey, leaving Hathath in the village of fear, a village that is only five days from Dream City, but even its inhabitants were not interested in trying to seek a better future.

After fighting a lot of beasts and living through strict weather and rough mountainous terrains; they made their ways to a point form where Dream City became visible.

This point was on the top of a hill that gave a clear few of this city.

Their eyes were filled with tears, and with the strength had left, they continued until arriving at the gates of Dream City.

"I wish Hathath had come with us," said Deroc in a low voice.

"Wow!" Joseph said. "Our dream has come true; we all have made our dream come true, together," said Joseph as he hugged Jonathan and Deroc.

"It was worth it, we did it!" said Deroc with tears in his eyes.

"That's right, we arrived," Jonathan said, his voice was full of enthusiasm.

"Now that we are already here, what are we going to do?" Deroc said.

"The first thing is to look where we can rest, then see where we look for work to cover our sustenance," instructed Jonathan.

"Yes, we will have to find some work for ourselves," said Joseph.

"I am looking for people who can work in house construction, can any of you help?" said a man who approached them with the curiosity to see if they were interested in working.

"Sure! I would love to," Joseph answered. "In fact, all three of us are interested." This man looked familiar.

"Is everything okay, young man?" asked the man.

"Yes, sir. We were just thinking how fast we can join you for work," Joseph tried to cover.

"So, follow me, I have a day's work and if you do it well, I can offer you more work," said the man.

The three of them followed. After a few hours, the man checked the work they had performed so far. He was impressed and thus offered them the job and a place to stay until they could stand by their own. Joseph was the one who had more experience in building houses. Because of that, the owner of the construction company sent him to study during the night while working for the day. He did such a good job that after four years of study and working with that man, he oversaw big projects. Everything Joseph did flourished. Then, the skills he had developed in his village helped him to land his position.

The three guys were living the life they wanted. Sometimes Joseph lost his focus to do other things that had nothing to do with his profession, but Jonathan was always there as a good friend; he never let him divert.

He was always by his side to advise him, counsel him, and tell him to "always think about the consequences of all the decisions you had to make in your life, because any bad decision could ruin your future." He would always remind him of the rough terrain all three of them had gone through to achieve this life.

Joseph maintained in his memory that the owner of the company was a familiar face, but he couldn't recall it.

Joseph would often think about the promises he had made to his grandpa Herald. He had fulfilled the promise of finding Dream City but he had yet to find his father. He would often cry at night as he would miss Hudson, Macy, and his grandpa.

His employer, Edgar, one day found him crying and asked him about it. He told him about his story and that he had promised his grandfather to find his father, but he has never done that.

"I fear the fact that I'm running out of time and I may never see my father again," said Joseph.

On this, Edgar told him about himself. "Dear young man, you want to know how I ended up here? I left the village Comforty when my wife died. Her dead devastated me. I didn't want to remember any of the memories that could make me miss my wife more. I even left behind a son who, during that time, was a child. I promised my father that I would return. I was so broken; I wanted to get away from that place. I changed my name."

"So why didn't you go back?"

"I want to. I dream of returning one day. I miss my son a lot. I left him with my father who promised me to take care of him until his last breath, but I never had time to go back. If I were to die right here, at this place, I would die with one regret—the regret of never meeting my son again."

By this time, Joseph had understood that this man was his father Hudson.

"You have already met your son, but you would never meet Grandpa Herald. Let me tell you that he has fulfilled his promise of looking after me in the best way and always misses you and Mom. He has been buried by Mom's side."

"Son, is that you?" cried Hudson.

"Yes, yes I am, Father."

The two of them hugged and cried.

This time, Joseph looked at the starry sky the same way he did when he was a kid, said, "Grandpa, I have fulfilled my promise," and smiled.

After several years working with his father, Joseph made the construction company become the most recognized in Dream City.

After many years living his dream, Joseph decided return to his village because he considered that now he was ready to rebuild the village Comforty and make it a place where its inhabitants could also dream. This was also one promise that he had made to his grandfather, so he had to fulfill it.

On his return, not only was his father with him but also Jonathan, his inseparable friend. Deroc and a few more pilgrims who came to Dream City from vast distances accompanied them.

They passed through the Village of Fear where they found Hathath again, and they stopped for an average of six months and helped rebuild that village. They taught its inhabitants not to be dominated by fear. Now that village was called the Village of Courage. Deroc was happy to see his brother again.

Many who accompanied them back to their villages knew another path that did not pass through the Valley of the Shadow of Death. This road was faster, and in a short time, they arrived at the village of Deroc and Hathath. Deroc stayed behind with his brother Hathath. He thanked Joseph and Jonathan for reuniting him with his brother. He told them he will never forget them and will come to visit their village.

There was a great celebration for Deroc and Hathath. For the inhabitants, it was an honor to have Joseph and Jonathan with them and now; they had brought another guest with them. This guest was Joseph's father. Now, with Deroc, the village was bound to improve. After many years working with Joseph and his father, Deroc had become a house builder and was all set to teach his brother Hathath and many people in the village the same work.

After leaving that village, Joseph, Hudson, and Jonathan reached the place where Joseph and Jonathan had met.

"Come with me so you know my village," said Joseph.

"No, you and I cannot go beyond this point together. It was a great honor to have accompanied you all these years," replied Jonathan.

"So, when do we meet again?" asked Joseph. He wanted Jonathan to come with him.

"I have always been with you, and I always will," said Jonathan with a smile.

"I do not understand. What do you mean when you said you've always been with me? I didn't get it," asked Joseph.

"Yes, that's what I said," said Jonathan.

"Have you not noticed that all these years, I have always been with you. I have been helping you no matter what. I have been advising you to avoid making bad decisions or getting away from the plan that destiny had already prepared for you. I have always kept you focused. I am your conscience, and I had to act in this way to always advise you and make sure every decision you took was the correct one. The destiny has kept a lot for all of us, and we need to make certain decisions to avail those bounties. I helped you in its way. Many people don't have one because sometimes either they get too greedy or too afraid to take risks; Deroc and Hathath can direct this example. Why didn't I have a place to go to? Why didn't I share my story with you, and why didn't I ask you about your name and where you came from?"

Joseph remembered all those moments when he had to decide with wisdom. With tears running down his cheeks, he said goodbye to his conscience in human form as it disappeared among the trees.

On his way back, Jonathan said, "Don't forget: I will always be there as I have always been."

People who had accompanied Joseph and Hudson embarked on their way to their respective villages while the two of them made their way to the village Comforty.

"Wow! This young man has surprised me. I never thought I would see him again," said Scarface as Joseph entered the village with Hudson.

Joseph and Hudson left the villagers speechless. A few of them came forward and applauded them for their journey back to the village.

Joseph took Hudson to Herald and Macy's grave and said, "Look, Grandpa, I kept my promise."

"The end," said Uncle P.

Everyone stayed silent. A few people began to look here and there. After sometime, Uncle P said, "For those who are still thinking, let me give you the message found in this story.

"The Village Comforty represents the place of conformity in the lives of some people. In fact, for most people, they prefer to keep struggling in the same place, even if they are not happy. They choose that way of life. Why do they do so? Because they are afraid of trying new opportunities that life offers them. This serves as a message for all of you: give yourself a chance; you deserve it.

"The fence of the village is the security zone. People like to live in this zone. Call it a secure zone or a comfort zone. Many people feel insecure about the thought of seeing themselves outside their zone of conformity and comfort. That's why they do not want to start a new job, a business, a new relationship, take the initiative, or move to another place, because they do not know how everything will turn out for them; they too fear the unknown. Thus, they prefer to stay in a state of struggle for the rest of their lives. Never forget: you are just one decision away from a different life."

Everyone listened.

This was the part where Sofia, Amanda, Oliver, Ariana, Jacob, Aidan, Ariel, and Emma could relate the most.

He continued, "The forest represents the unknown that every person must face. None of us know what the future holds, but to explore our opportunities, we all must allow ourselves to take those chances and explore the tendencies and the potentials of the unknown. Many are afraid to leave their place of conformity and undertake something new in life; they fear the unknown, and above all, they fear failure. They are afraid of the outcome that may come with the decision. You will never know the result if you do not try. Dare yourself to dream and take a leap of faith. If you aren't brave enough to decide, you will end up on your deathbed with a lot of regrets. Just like Herald."

On this, Amanda looked at Jeffrey and smiled.

"Scarface represents those who have failed in their attempt to achieve something in life. These are the people who let their own experience shape other people's life choices. Since they encountered failure, they cannot see other people succeed. Never forget it is upon us to write a good ending to our stories; life is our canvas. We can paint it the way we like. When people like Scarface see people who are determined to pursue their goals, they try to discourage them with negative words. They try to stop them. The sad thing is that a vast majority end up living under the opinions and comments of what other people say about them. The second majority are those who are like Scarface who have failed. Such people make others bury their dreams."

At this, Aidan and Jacob looked at each other, but then, Aidan averted his gaze.

"A great majority of the inhabitants of the village represent those who have never tried to achieve anything in life, and as they have never done it before, they see other people who deal with their senses and envy them. This is because they do not believe in themselves and they want others to end up as they did."

Ariana could relate them to the people in her hospital.

"The village of Deroc and Hathath represent a stage in the life of many people who on their way to success will have to stop and help others to get up. It teaches us something important about being a good human being. While we try to achieve our goals in life, we should always be ready to help others; this show if we have enough human left in us or not and confirms if we are living or breathing. Never forget, there is a different between people who live and the people who breathe; someone who is living will care about how others around him are doing while someone who is only breathing wouldn't.

"Deroc represents those people who, on their way to success, are not willing to help others who have fallen. Such people are only breathing and not living. However, Deroc soon began to live after he met Joseph and Jonathan. This is where he began to help others and began to feel for others; once people feel for others is when they live.

"Hathath represents the ones who are always afraid to take risks in their lives; they let the time do nothing. These are the ones who with regrets. Mark my words: time flies and the time never wait for us. It will go by and we will not know. It is always now or never. If you have something in your mind, if you want to educate the masses, if you want to put your ideas forward—speak up now; do not wait!

Sometimes, they start well, but on the way, the 'unknown' fills them with fear. This fear of unknown stops them from progressing. They prefer not to continue the journey and stop. These people regret much more than those who had never started the journey of pursuing their dreams. Why is this so? Because they have lived through both, they know how it feels to take steps toward success and they also know how it feels to fail and to stop; they are in the worst position."

Then Uncle P got up from his chair and said, "My dear audience! Do not give up; you are closer to your destiny than you can imagine. Maybe your success is around the corner, too close. Do not stop. Do not fear what is coming. This may stop you from writing your success story just by an inch. Don't let the fear take over you. Don't do this injustice to yourselves.

"The Valley of the Shadow of Death represents the place where many dreams have been buried. We all can relate to it. We all must have left something in our lives. For whatever reason, we all have at least one dream that we had to bury—teach your kids something different, teach them to pursue their dreams, teach them to not kill their dreams. There are people who died with their dreams. Don't be one of them and don't let your children become one of them. Do not let your dreams go to the grave with you. Give life to your dreams.

"Jonathan represents consciousness, as I have told you already; your conscience is the one that corrects you when you are wrong and encourages you when you are doing the right thing. Let it guide you right; don't suppress it under ego, fear, frustration, or overconfidence. Our conscience helps us to make wise decisions; sometimes when we do not want to do it and want to surrender, our conscience speaks to us and tells us to do otherwise, listen to it, let it guide you. Take yourselves to where the road is taking you."

"Joseph represents all other people—in fact, the minority who triumph, they represent the ones who do not pay heed to what the masses say ... they are the ones who create their own ways and believe life will be how they make it; these are the people who win.

"As an ending note, I would say that you already came to this world well equipped. God has provided you the might and many to make it through every hurdle and cross every barrier in life. God has given you a brain that will help you tackle anything that life puts in front of you. You are no less than others around you. God loves all of you. You only need to understand that it is the fear that holds you back. If not anything else, people fear that God loves them less; there is always some kind of fear holding them back. Do not let your fear be greater than your dreams. I challenge you to dream. I challenge you to make a life that you love. I challenge you to live, and I challenge you to live a regret-free life! Thank you very much, everyone." These were the final words of Uncle P.

With this, the audience broke their silence with a round of applause and stood up in honor for Uncle P as he left the stage.

9

Proposals

After Sofia and Oliver listened to the great story told by Uncle P, they realized how important it is to change the perspective of looking at things in this life. Both of them were sure of two things: one, they needed to move forward, and two, they were perfect for each other.

We often think that moving forward means leaving the people in our past behind, but this is not the case. We could never leave the one we love behind. They always stay in our hearts. Just like parents who lost a child and then bore a new one, does it mean they have replaced the child who died? No, they didn't; they didn't give his place to the other child, as this is impossible.

The same is the case when someone loses a spouse or anyone they love; they move on, but they do not replace the people lost.

Sofia and Oliver both understood this. They began to treat each other differently. Oliver now knew Sofia was the woman he wanted to spend the rest of his life with. She was the one who would love him and his children unconditionally. She was there all along; he was too busy or too naïve to see it.

Sofia understood everything that happens doesn't lead to losses.

We may not know why we are being inflicted with adversities, but we will see how they turn into something viable and good. A human being is so impatient that he tries to find reasons and wants to see results instantly. We forget that patience never goes unrewarded. Even in the darkest of times, we realize, only if we know how to turn it on.

We need to stay at peace when God makes things work His way. None of us know what our future holds, but He does.

Sofia had found a man who loved and respected her. She wanted kids desperately, and God had rewarded her with two beautiful kids who reciprocated her feelings for them.

Sofia and Oliver didn't talk much during their way back to Sofia's place. However, the air was losing the ease it had. Both of them had stayed silent for long. Sofia was looking outside the car and smiling to herself.

"Do you mind if I ask what is making you smile right now?" Oliver broke the silence.

"What? No, I am not smiling ..." said Sofia while blushing.

"You were smiling, I saw you," said Oliver.

Sofia smiled.

"Now tell me, why you were smiling?" asked Oliver.

Sofia still smiling said, "Why? Don't you like to see me smile?"

"Yes, I do," said Oliver.

Both were taking steps toward each other. As happy as they were to do so, they were also frightened. They were thinking both ways; they couldn't let such a worthwhile life partner go, nor could they rush things too much that they lose their charms.

They stopped at a traffic signal. Another car stopped by Sofia's side. To skip the awkward course Sofia and Oliver's conversation had been taking, Sofia looked out her side. In the car was a family of three: a child in his car seat and a wife talking to her husband while driving.

Sofia felt like she was looking at herself in a mirror where Carlos was driving, she was sitting next to him, and her adopted child was sitting in the back seat.

"What are you looking at?" asked Oliver.

"The family," replied Sofia.

"What about them?" asked Oliver.

Sofia turned toward him with lots of question marks on her face. "Like, you know ..."

"No, I do not know. You tell me," replied Oliver.

"You know. Or don't you?" replied Sofia. She was becoming defensive.

"I do not; therefore, you need to tell me, Sofia."

"Look, I don't understand why you are doing this Oliver, but I know one thing, you know why I was watching them."

The signal turned green and Oliver didn't have time to continue. He began to drive. Sofia let out a sigh of relief that this topic would not be pursued any further. However, she didn't expect what happened next.

Oliver stopped the car outside an ice cream parlor. Then he said, "Which ice cream would you like?"

"Umm, are we going to stop for ice cream?"

"Not always, but only when we need to. Now, tell me, which ice cream flavor would you like to have?"

"Butter pecan, my favorite!" said Sofia in excitement.

Oliver could see the excitement of an innocent young girl in her eyes. He smiled and Sofia began to blush. Then Oliver added, "I never have watched you closely, and now, I have realized that you are a beautiful woman."

Sofia didn't answer, and she began to look down. Oliver went to buy the ice cream. Oliver came back with butter pecan ice cream for Sofia and vanilla for himself. Sofia began to eat excitedly. Oliver saw how innocently she savored the ice cream. She was not that middle-aged woman who would take care of his kids. She was a young girl this time. She was an innocent girl who was too delicate to be exposed to the harshness of this world. She wasn't made to face the harsh realities of life. She was someone who needed protection, who needed security, and who needed love. As Sofia realized she was continuously being watched, she looked up and said, "Um, what are you looking at?"

"You," replied Oliver.

"Umm! I know, but why?" asked Sofia.

"Just watching you eating ice cream," said Oliver.

"Okay!" responded Sofia with wonder.

"One more thing," said Oliver.

"Now what?" asked Sofia with concern.

"About why were you looking at that family in the car by your side." Oliver quickly came to the point.

"Please don't bring this up," pleaded Sofia.

"Why?"

Sofia didn't reply.

Oliver understood she kept something in her heart that had burned her for a long time. Then Oliver held her hands. This was when Sofia broke into tears. "Why did you bring this up?" She cried.

"Were you able to put it to rest?" asked Oliver.

"No! No, I wasn't." Sofia burst into tears. "Every time I try to put it to rest, it just doesn't happen. I was watching how blessed that woman was. She has all the bounties of this world. This could have been me, this could have been me, and this could have been me!"

With this, Oliver moved forward and pulled her close for a hug.

"Do not cry Sofia, do not. We do not have the answers to all things that happen in this world, but I'm sure of something.

Through all this, something beautiful will emerge. I don't want you to carry this with you throughout your life, so I want you to pour out your heart."

Sofia held his shirt as she cried.

"I was so close to get all of this, very close. Carlos and I had gotten into an agreement to adopt a kid and look what happened. A drunk driver destroyed my entire world in a moment and has left me alone. You don't know how desperate I was to give a child to Carlos, but God didn't give me a chance. Why did it have to be me?" cried Sofia.

Oliver wiped her tears and said, "Look, Sofia, you will soon get to know that whatever happens has a reason behind it. God has greater plans for all of us. Carlos's death is a huge loss, and no one can ever replace him in your life. I totally understand that, but you will have to look forward and you will have to move on."

As Sofia heard this, she got up and began to wipe her tears. Then she said, "I want to live a happy life."

"I know you want to, and you will. Let's drive you home," said Oliver.

"Yes, please," said Sofia. Sofia's eyes were swollen and red rimmed.

As they reached outside her apartment, Sofia said, "I took a lot of your time, Oliver. Thank you for everything."

"No, thank you. I must admit, it was a good time. Thanks for taking me out of my comfort zone," he replied.

Then she added, "You are a good man, and an excellent father … Any good woman would like to have a man like you as a husband …"

Sofia came out of the car, and so Oliver. He accompanied Sofia to the entrance of her apartment on the first floor. Before she went inside her apartment, she kissed him on the cheek. He blushed. Then he returned to his car and put it in gear.

In the morning, Oliver was in the kitchen making coffee. Anthony and Leslie were at Betty's house. Suddenly, he heard the bell at the door. It was Sofia.

"I hope you haven't started to prepare the coffee," said Sofia.

"I'm trying. Why?" asked Oliver.

"Because I would like to prepare coffee for you. You should sit," said Sofia as she signaled him to wait. Sofia began to prepare coffee for both of them and Oliver sat on the kitchen counter.

"You know this art well," said Oliver as he smiled.

"Elizabeth would always ..." Then Oliver paused. He didn't finish.

"This is your turn now," said Sofia as she held his hand.

"I do not know what to say," replied Oliver.

"You know. I would like to hear your story!"

As Sofia said this, she took Oliver to the family room. Both of them sat on the sofa and then Sofia continued. "So yes, Mr. Lawyer, I am all set to listen. You were talking about your wife, Elizabeth. Come on. Go ahead, please," said Sofia in persuasion.

"Okay, so Elizabeth would always ask me what I would do if she was not around." Oliver broke in to tears. "I used to tell her that would not happen, and that I would see her making breakfast in my kitchen until I take my last breath." Oliver held Sofia's hands and said, "You know, she would always ask me, 'What if I close my eyes before you?' I would always be afraid of this thought. I would silently pray to God that He never brings this day. I was so sure that nothing like that could ever happen. The day I had met Elizabeth, I thought, this is the woman I will spend the rest of my life with, and when she was diagnosed with cancer, I couldn't believe this was happening to the love of my life. Leslie was so young. I was broke when the doctor told me the cancer was increasing day by day. It was eating my love from inside. She lost a lot of hair; that one day, I saw her tense. She would burst into tears. All her hair was falling out, and each time she would comb her hair, more strands would come out. One day, she got up and told me she wanted to shave her head."

At this, Sofia too began to cry.

"Oh God," said Sofia.

"Yeah." Oliver too was in tears.

"You know, I shaved her head—her beautiful hair. She was dying; my love was dying," he sniffed.

Then he continued, "I could never believe she would leave me soon. Every day when she would go to bed, I would move close to her while she would be asleep and listen to her heart thumping and her slow breaths. I would always be afraid that it could be any moment when I try to feel her breaths and I could not feel any."

Tears fell down Sofia's cheeks and on to Oliver's hands, which rested in hers.

"Leslie would ask me why her mom is always wearing a scarf, or why she couldn't run with her in the park, and why she often visited the doctor. I had no answers. I could not tell her that her mom would not stay with us for long."

"Oh, Oliver, even you have been through much, but you also had never let it out," said Sofia. She pulled Oliver into a hug.

Oliver continued. "I am too afraid of the fact that people around me will leave soon. I cannot afford the loss of loved ones. I am just not made for it."

"You will never have to—not anymore," replied Sofia.

Both of them sat on the sofa throughout the morning and talked their hearts out.

A week later, Oliver took the kids and Sofia to the cemetery. Sofia had a bouquet in her hands. She placed the flowers on the tombstone and stepped aside. She could see the tears in Oliver's eyes.

Leslie, who was holding Oliver's hand, went forward and sat by the tombstone. Anthony accompanied her. "We love you, Mom," said Leslie.

"We will always remember you. I thank God for giving me a

mother like you even though we could not share for a long, but I know you are in heaven now," said Anthony.

"Grandma says I have a smile just like yours, and that make me happy because I have something of you," added Leslie.

Oliver and Sofia looked at each other. Though the kids were young, they had spoken something which would have made their mother cry if she was alive.

"Get up my babies. Your mother must be thrilled and honored with these words," said Oliver. He knew the kids would soon break into tears, so he had to divert their minds elsewhere.

"Take them to the car, Sofia, and wait for me there," added Oliver.

Sofia nodded and took the kids with her.

Then Oliver sat by the tombstone. He too had tears in his eyes. "Did you see how affectionate our kids are? They have a heart just like you. Thank you so much for being my wife, my friend, and my counselor. You were the best thing that had happened to me. God had blessed me with you. I see you in Leslie daily. My beloved and my beautiful wife, I always loved you ... I will never forget you and I will always keep you in my memory," said Oliver.

He wiped his tears as they dropped one after another, got up and went back to his car. He sat inside, but no one said a word.

They soon reached the house. The kids got out of the car on their own, Sofia, and so did Oliver.

"These kids are growing up so quickly," said Sofia to break the ice.

"Yes, they are."

Before they went inside the house, Oliver stopped Sofia. "Would you mind having dinner with us tonight?"

"Um ..."

"Please?"

"Okay, I will," replied Sofia.

In the evening, Oliver's mother also joined them for dinner. Some colleagues, close friends, and Oliver also invited a few distant family members for dinner.

"Umm, I thought this was a casual dinner. Is today something special?" asked Sofia.

"You will know soon," replied Oliver with a broad smile.

On that night, Oliver asked Sofia to marry him. In front of everyone, he got on his knees and said, "Sofia, I have fallen for you. Will you be my wife?"

Sofia got nervous and didn't know what to say.

"Say yes," said Betty.

"Yes, I will," replied Sofia, who was surprised. She too had developed feelings for Oliver, but she didn't expect all of this to happen so soon. Sofia and Oliver hugged. Everyone clapped, and the kids were happy. Betty was in tears. She wanted the best for both of them and her grandchildren. It was like a dream come true for her. Anthony and Leslie came forward and hugged both of them.

"Our family is complete! Yay!" said Leslie.

Oliver and Sofia kissed her on each check. One of Oliver's friends clicked a picture of this beautiful moment.

Three months later, Oliver and Sofia celebrated their outdoor wedding in a park by a beautiful lake. The journey from Mr. Anderson to Oliver had been a difficult one, but worth it.

***

After hearing the story of Uncle P, Ariel and his parents left the school building with everyone else. They saw their coach Jengo in the parking lot.

"Greetings, sir!" said Ariel.

"Oh hello Ariel, you are here?" replied Jengo.

"Yeah, the show was worth it. How we could miss such a show?"

said Peter from behind as he came forward to shake hands with Coach Jengo.

"I need to rush home. My wife is waiting for me. I have to take her out for dinner," added Jengo.

"Oh yes, please give my greetings to her," said Jessica as she stepped forward.

"Yes, sure," said Jengo.

"Also Ariel, I would like to see you tomorrow at ten in the morning," added Jengo.

To this, Ariel had question marks on his face and so did his parents.

"For the basketball practice, young man!" said the coach as he patted him on his back.

"Okay," replied Ariel. He didn't show much enthusiasm. He was keen to miss the practice. He thought his classmates would still make fun of him.

Jengo sat in his car and off he drove. So did Ariel with his parents.

"Are you ready for the practice, sweetie?" asked Jessica during the car journey.

"I don't think so. Albert and Joel will make fun of me again. I don't want that," replied Ariel.

"They would not, my boy. Give yourself another chance," said Peter.

"Yes, we will see," replied Ariel.

They didn't talk much during the journey. As they got home, Ariel rushed to his room.

"Wait, my son," said Jessica.

"Yes, Mom?"

"I want to say something. It is your life and your decision will be final, but don't forget, you are always one decision away from a different life," said Jessica.

Ariel didn't reply. He just listened. Jessica planted a kiss on his cheek and said, "I will not pressurize you anymore. It is your call now. Good night."

"Good night Mom." With this, Ariel went to bed and slept.

The next morning, it was nearly eight in the morning. Jessica and Peter were having their morning tea.

"Do you think he would go?" asked Peter.

"I don't know. He was unmotivated in every way. I would not blame him if he doesn't," replied Jessica.

Soon, they heard someone coming down the stairs. It was Ariel. He was all set with his bag and was wearing his jersey. "Dad, I need to be on time today," said Ariel.

"My boy? Are you going for the practice?" asked Jessica in awe.

"Yes Mom, as you said, I am just one decision away from a different life, so I have made my decision. I am not quitting," replied Ariel.

"I am proud of you, my son!" said Peter. He got up and hugged him.

"Just give us five minutes; we will get ready and go together to drop you for practice," replied Jessica.

"Okay. I will wait," replied Ariel as Jessica and Peter rushed to get ready. Ariel had different confidence that day. His face was beaming.

He had decided he would let no one put him down and prove his enemies wrong.

As Ariel reached the court, everyone clapped for him and his performance. Even Albert and Joel came and apologized to him for their misconduct. They must have heard from their mother.

"We are very sorry, Ariel," said Albert.

"We shouldn't have hurt you like that. We need you on this team," added Joel.

"It's okay, guys. No hard feelings. From now on, we need to make sure that the team makes it to the finals and wins it," replied Ariel.

"Yeah!" said Albert and Joel together.

With this, Albert and Joel gave a high-five to Ariel and all of them joined the practice together.

The following year, the basketball season began, and this time, the school indoor basketball court had a bigger audience than last year. Both the teams, the Jefferson Jaguars and the Roosevelt Rockets, played spectacularly. Except this time, they are in the state finals.

Ariel has the ball in his hand. Many players of the opponent team come in his way, but he safely takes the ball toward the basket.

The audience jumps on their feet. It was the same scene from last year. Ariel threw the ball. Jessica and Peter prayed in the stands. Ariel's hard work and consistency pay off. Ariel scored the winning goal and closed the season. He made his father, his mother, and everyone who believed in him proud. If Ariel had decided otherwise and hadn't given himself a chance, could he actually bag the trophy under his captainship?

There could have been anyone in place of him, and it may not have mattered to him if the Jefferson Jaguars had ever won the tournament or not, but could he feel the same confidence in himself like he did after giving himself a chance?

What most don't understand is, we need to give ourselves chances. We are our last and our best hopes. Be all ears, but don't let what the foolish say get to your brain. If we follow what the fear says, we should also be ready to end up where the fear is nowhere.

There are people who would demotivate us. It may not always be because they do not want to see us succeed, but it is just in their approach. They cannot think beyond a certain point; they cannot imagine progressing beyond a certain point.

They are the ones who give up too soon. Are you like them? This is upon you to decide. Why do you think you cannot do it? Why do you fear showing what you are good at? Ariel was a young boy who pursued his dreams. It is still not late for you to do so, regardless of where in life you are and regardless of your age. It is never too late.

10

My Land's Calling

Everyone left the event hall on a positive note. Uncle P's fable motivated people to look forward and explore new opportunities in life. Ariana, like everyone, came out with a smile on her face. She needed this kind of cleansing. She needed this advice; she needed someone who could push her to stop fearing change. She now understood things happen for a reason. She had a clear mind to think about the opportunities that life had provided her.

The sisters and Lily sat in the car. Ariana began to drive.

"It was a good show. I loved it," said Laura.

"Yes, it was," said Ariana. She was smiling throughout.

Lily had fallen asleep in the backseat. It was already late. Ariana and Laura were also tired, but all of them had a good day. Ariana kept making scenarios throughout the journey in her head about the repercussions that she would have to face if she took a decision to leave her current job.

The sisters didn't talk much during the journey. As they reached home, they bid goodnight to each other and go to bed.

Ariana had to leave the following day, which was a Saturday. She got up during the night. She changed sides, but she wasn't able to sleep. Suddenly, she heard a noise coming from downstairs. She got up, put her rope on, and went downstairs. She made her way to the kitchen through the dark hallway. Someone was in the kitchen. She heard the refrigerator open. As she stepped inside the kitchen, she saw Laura trying to find something in the refrigerator.

Laura also heard the footsteps. She carefully got up and turned.

"Ah!" said both of them.

"You nearly scared me, Ariana! Oh, God, I'm pregnant!" exclaimed Laura.

"I heard some noise, so I came down, I am sorry ... I couldn't sleep," replied Ariana.

"What? Well, I also wasn't able to sleep for a long, then I began to crave ice cream, and look, here I am," said Laura as she laughed with her hands in the air.

"Take a seat. I will get the ice cream," said Ariana.

Laura took a seat and Ariana took out a tub of ice cream from the refrigerator. Laura delved into its sweetness. She enjoyed every bit. Ariana looked at her and smiled.

"Ah, my sister, you will leave in less than fifteen hours," said Laura as she called her for a hug with the motion of her hands.

"I have to, Laura. I have decided nothing, but I am surely going right now."

"I will miss you, baby sister. I hope that you soon come back and live with me; you know, a family always needs to stay together. We are the only ones left for each other in the world," said Laura.

"I know, Laura."

"You know, before mom died, she told me to take care of you, and do whatever I have to for you, but how will I do that if I get to see you once every three years?" asked Laura.

"Sister, I am a grown-up girl. I know you care about me, but we sometimes cannot do much about the situations we are in," replied Ariana.

"We can. Why not? You can come and live here," said Laura.

"Okay, okay! I will think about this, but for now, go to bed," replied Ariana.

"Yeah, I know! You are trying to avoid this topic. But don't forget: I will call you daily and bug you for this. Ha!" said Laura as she got up.

Both the sisters made their way out of the kitchen. Laura went toward her room, and Ariana went upstairs.

"It was nice talking to you, Laura," Ariana called out from the stairs.

"Same here, babe." Laura's voice could be heard from a distance.

Ariana went to bed with much more positivity. She had packed everything in place.

In the morning, everyone in the house woke up with the sun. It was a rather emotional day. Laura, who was expecting, wept throughout. Ariana tried to console her and told her she would visit her soon. Lily, like her mother, cried and pleaded before Ariana to stay.

"She will visit us soon, and may even live with us," said Henry as he wiped Lily's tears.

"Promise?" said Lily.

On this, Henry was quiet. He couldn't make a promise for something he himself wasn't sure about.

"Let's hope for the best, my dear," said Ariana from behind. Lily hugged her tight. She would soon leave for the airport.

Ariana returned home Saturday afternoon. While on the plane, she was thinking about new changes in her life. She was ready to welcome new things. The plane landed at the Orlando International Airport. Ariana took a cab back to her apartment. As she reached the apartment, she texted Laura.

"Hey sis. I hope you are well. I had a good time. This was something that I needed. I have reached home safely. Love, Ariana."

Ariana took a shower after getting home, prepared herself a coffee, and then called Margaret.

"Hey, Margaret. I am back," said Ariana with much enthusiasm in her voice.

"I am glad," replied Margaret.

"I have so much to tell you," said Ariana.

"Then go ahead; what are you waiting for?" said Margaret. She too wanted to know how Ariana felt now.

"You had an interview. Tell me how it was," said Ariana.

"Umm, I will start working for this new hospital, Ariana. The package and the fringe benefits are much better than what I am getting here."

"That is fantastic!" said Ariana. She was happy Margaret had a new job, but that she would leave soon made her sad simultaneously. Ariana had mixed feelings.

"It is my last week at Jessie Hospital, Ariana; you have been such a great companion for me," said Margaret.

"Margaret, let's not discuss this over the phone. I am thrilled for you and I am sad at the same time; you are among the few positive people at work. I will miss seeing you so much! I am already in tears," said Ariana.

"Oh Ariana, you are literally my best friend!" replied Margaret. "I will get back to you, but I am running late for work," added Margaret.

"Sure, take care. See you soon!" said Ariana.

"See you," replied Margaret, and hung up the phone.

Ariana then sipped her coffee and began to ponder over the story of Uncle P. She remembered the job opportunity that came forward in the Maryland hospital. She then got her laptop, opened it, and typed the hospital's website into the search engine. She had memorized it.

"Your best friend will not work with you anymore, Ariana," Ariana said to herself.

She checked the positions at the Maryland hospital and began to think about the life that she could have.

Can I endure all the discrimination at work? Thought Ariana.

Though Dr. Judith, the supervisor assistant at Jesse Hospital, would always support her, could she stand against all the discrimination and hate by other employees? This concern immersed Ariana in deep thinking and evaluation. She had to decide; back in Maryland, her sister was waiting. However, this was her first workplace, and she had stayed there ever since.

"May God tell me to what it deems fit for me," said Ariana as she closed her eyes.

Then she opened them and started filling out the application. She uploaded the documents and submitted.

"This was it," said Ariana. She had taken her step. All that she could do now was to wait. She closed her eyes again and prayed to God, "Oh God, if this is good for me, then lead me to that place; if it is not, then change my mind, you know what is best for me." With this, she continued with her routine.

A week passed, and she received nothing. For the next month, Ariana checked her emails again and again. Laura would also call her to know if she has changed her mind. Ariana always told her she hasn't changed her mind because she had received no reply from the Maryland hospital.

"I'm three months pregnant, Ariana. I want you to be around when I deliver the baby," said Laura.

"Even if I don't move to Maryland, I will come to you when the baby is due. You take good care of yourself and your baby," replied Ariana over the phone call.

Ariana would always think to herself, what if I never get a reply? Is staying here worth it?

After four months of desperately waiting; an email notification appeared on Ariana's desktop. It was from HR MARYLAND HOSPITAL. They had email Ariana for an interview in three weeks. Ariana rejoiced with this news. She wanted to call Laura right away, but then she surprised her. She called Margaret and told her about the email. Even Margaret was happy for her.

Ariana landed at the Baltimore airport two days before the interview. She knocked at Laura's with her suitcase. Only Henry knew about the interview. Laura opened the door, and that surprise overjoyed her to see her younger sister. Ariana too had tears in her eyes. Laura had a bigger baby bump. Both the sisters hugged each other.

Two days later, Ariana went to the interview and was extended a job offer the next day. Ariana understood it is always worth waiting. This job opportunity had reunited her with her family after a long time.

What most people don't understand is that it is important to stay close to the people we love—whether in our family, or someone not related to us by blood. The main idea is to stay close to those who want us and need us in their lives. It is important to feel loved. If something destroys your peace of mind, then it is not worth it. Looking for a better future, a better job, or a better career, it is always important to keep your peace of mind and people you love as a priority. One should never fear starting again. Turning the page is important. Going against the flow is important, and taking steps is important, for we do not fear change, but we fear the unknown.

Was it so hard to understand for Ariana that the loneliness was taking a toll on her? Was it so difficult to understand for her that the workplace was taking away her happiness?

No, it wasn't—she knew all of it, but she feared to take a step.

When the nine months arrived, Ariana was the second person to hold Laura's second child. It was a baby boy and Ariana was the one who gave him his name. He was named Adriel.

11

The Guarantor

After hearing the soul awakening-story of Uncle P, Jacob, like every other listener, was moved. The story had rekindled his motivation and spirit. He was clear about a few things. First, everyone has different opinions and outlooks to life. We all need to accept those. Second, even our family members can disagree with our approaches to life and careers. Third, no matter which path we choose, we will always have to fight through the barriers and the hurdles that the path brings along, for only the difficult roads lead to beautiful destinations.

He knew even if he went forward with his idea, got the loan, and started his own restaurant venture; the initiative would only increase his challenges. It would not be easy, and it would increase the pressure on him. With this, he also knew if he doesn't take the initiative and continue with his paid employment, he would regret it for the rest of his life.

His father, Aidan, also didn't talk on his way to the car. The father and son sat in the car and Jacob began to drive.

Aidan looked like he was in deep thinking.

"Are you okay?" asked Jacob to break the ice.

"What?" replied Aidan as if he had just been woken up from a deep sleep. "What did you say?"

"You have been quiet since we left the show. Also, I noticed you are thinking about something, what is it? Are you okay?" Jacob asked again.

"Yes, my son, I am okay. I was just thinking about the story of Uncle P. I hope my silence doesn't bother you," replied Aidan.

Jacob smiled to his father and then continued to drive. He couldn't help but smile throughout the journey. The story had been soul-awakening for him.

"Dad, your silence doesn't bother me. That you accompanied me to the show has made me happy!" said Jacob as he drove.

"No, my son, I should be thankful to you for taking me to such an incredible show. We all need to hear these kinds of stories and such words in our lives," replied Aidan.

"I am glad, Dad."

Jacob's car was outside Aidan's house.

"Take good care of yourself, my son," said Aidan before leaving the car.

"We should take some time and talk about the show," added Aidan.

"Yes. Why not? I would love to," replied Jacob.

"Come over for breakfast tomorrow, or maybe we can talk over a coffee? What do you think?" asked Aidan. The way he impressed a meeting clarified that he wanted to talk.

"Um, Dad, I would love to, but tomorrow isn't possible for me," said Jacob.

"Oh, okay," replied Aidan. With this, he finally stepped out of the car. "Good bye, son," said Aidan.

"Good bye, Dad," replied Jacob.

Jacob made sure that his father got inside the house and then drove away.

Monday arrived and Jacob was all set to visit the bank. He looked at his phone. His father neither called nor sent him any kind of message to wish him the best. He dressed a suit that his father had gifted him to wear on his first day of work. He considered it to be a special suit for him. He looked at himself in front of the mirror, grabbed his car keys, and headed out. He turned the car on and drove to the bank. He kept saying prayers throughout the ride and also checked his phone, like he was expecting a call from his father.

Jacob arrived at the bank. He parked his car in the bank parking lot. He had his proposals ready. The loan process was a twofold procedure. The first part was an interview with the bank manager. Jacob had already cleared the interview. This time, the team had to review his proposals and decide if he was worthy of a loan.

"Mr. Miller," said the manager.

"Yes, sir," replied Jacob.

"I hope you have your proposals and the documents in place."

"Yes, I do." With this, Jacob handed him the files.

"You may take a seat and wait for our financial analysts to study the proposals," instructed the manager.

Jacob took the vacant seat in the hallway. He constantly checked his phone and looked at the entrance, like he was expecting someone to come. After thirty minutes, the bank manager came out. Jacob's anxiousness was peaking during this time.

"You may come with me, sir," said the manager.

Jacob nodded and stood up. He followed the manager into the meeting room. A team of analysts were at the other side of the table. The manager asked Jacob to take a seat. He too sat on one of the empty seats across the table.

"Thank you for your patience. We are pleased with your proposals. We can see potential in it," said the bank manager.

Jacob passed a faint smile.

"However, we have been discussing the risks we have seen in your project. As a financial institution, we need to take care of the gearing that the project may have."

"Yes," added Jacob.

"To consider your project further, it would not be worthwhile for our institution to take that risk, because until now you do could not put something as collateral."

"Collateral?" Jacob repeated.

"Um, yes. Collateral so we know that in case your results are not per the proposal and we know where we can get the money back," continued the bank manager.

"I am afraid I may have nothing to offer as collateral," Jacob said.

"Oh, we have seen that you have all your documents in order; you have excellent credit. We have noticed that you are a good administrator and you can do a good job, but what if things don't go well? Businesses are risky. The returns are greater, but so are the risks. We need to insure ourselves."

"I understand." Jacob's throat had already dried up. He knew they had rejected him and thought his father was right—it was better for him to stay in paid employment.

"As much as we want to help you, I am afraid our hands are tied. This time, we are short of something that we can hold on your part so that our institution feels safe and can trust you with the funds. If you have someone who can give you a second signature or has a house or a business that can collateral, then we'd be willing to give you the loan. Do you have someone who can do it?" asked the manager.

Jacob was about to say no. He thought his dreams have been shattered. He could not pursue them. He loosened his tie a bit. He already felt hot in that air-conditioned room.

Suddenly someone entered the room and said, "I am sorry that I kept on hearing the entire conversation, but I can be the guarantor for the loan."

Jacob turned around to see who was there. The voice was familiar, but he had to make sure. The manager and the team of analysts also began to look at the man who had just meddled in the meeting.

"Excuse me, but who are you?" asked the bank manager.

A huge smile appeared on Jacob's face and his eyes had tears.

The man said, "I am Aidan Miller, Jacob's father. I can be the guarantor."

Aidan surprised Jacob.

"Dad ..."

"So are you willing to become a guarantor for your son?" asked the manager.

"Yes, I am. I have complete faith in him," replied Aidan.

"Okay, Mr. Jacob, are you willing to accept him as a guarantor?" asked the manager.

"He is!" added Aidan as he patted Jacob on his back.

The bank prepared the documents to be signed. Jacob couldn't believe his father came in to help. Right on time, when he needed it most. He knew that God had answered his prayers, his father had believed him and he was so close to pursuing his dreams.

Jacob, Aidan, and the bank manager signed the documents. As Jacob and Aidan made their ways out of the bank, they didn't say a word.

Once they had stepped out of the bank, Jacob said, "Dad, you don't know the favor you have done for me. I literally will forever owe you for this." He held Aidan's hands and said, "Thank you, Dad. You have always been the best. I love you."

Aidan hugged him and said, "No, my son. You need not thank me. I believe in you and I know you will be a successful entrepreneur."

"I am the happiest son on the face of this earth, Dad!" exclaimed Jacob.

"My son, I would like to apologize to you. I have been so unwelcoming with your idea, but I feared failure. What if you end up like me? You would be heartbroken. This is your dream. You will be hit badly. I cannot see you hurt. Therefore, I didn't support your idea."

"Then what changed your mind, Dad?" asked Jacob.

"Your conqueror spirit and the story of Uncle P. I have understood we need to give ourselves another chance. If I could not succeed, then I should make sure you do. Why should our kids be guided by our fears? No, you don't deserve to live this kind of limited life. I will support you through and through, my son. Just do your best!" impressed Aidan.

"I am sorry, Dad. I should also apologize. I have been so rude to you. Only if I could have understood your point, I wouldn't have been so naïve. Thank you, Father, for everything! I will need your counseling and support to run the restaurant."

"I will always support you, my son. You had your repercussions, and all we needed was to stop imposing our thoughts and start listening," said Aidan. He then hugged his son.

"I want to show you the restaurant model that I want to open. To give you an idea, it's a social restaurant where people can buy food or something to eat and also have access to the Internet. It will have the comfort of home. You can even watch movies in the restaurant if you pay a certain amount," said Jacob.

"This is a good idea, my son. Your ideas are driven by the customer-centric approach. Such ideas always work. I feel so proud to be your father," said Aidan. His eyes had brightened.

Jacob continued, "This restaurant model will make people feel like home. They can enjoy a moment with friends, families, or maybe alone. I wish to create something that will attract mostly young people and, in particular, students who wish to take a moment to recreate themselves. Best for me time."

"I like your ideas, son," said Aidan.

Both Jacob and Aidan walked toward the parking lot. They hugged each other.

"Call me if you need any help," said Aidan.

"Will do!" said Jacob as he waved good bye. Both got in their cars and went home.

On his way back to his house, Aidan kept thinking about the ideas that Jacob had proposed. He thought to himself that his son had grown up and is on his feet. He admired his new endeavor and was also thinking about his failure in the restaurant business.

As the car stopped on the traffic signal, Aidan went into deep thinking. He thought about what went wrong during his time, and he soon realized Jacob was right—he didn't notice when the market changed, and soon, he was left out to dry. The scene of his restaurant being foreclosed on began to play in front of his eyes, and soon, loud honking brought him back to his senses. The traffic signal had turned green, and he was stopping the traffic. He looked out of the car window, said sorry, and started driving.

Aidan soon reached home. Upon reaching home, he saw a missed call from Jacob. He called him back, but Jacob didn't answer. "Sorry, I missed your call, son. Call me back once you are free. Love, Dad."

Soon, he heard a knock on the door. He went to see who was there. It was Jacob. He opened the door and said, "Son?"

"Yes, Dad. As soon as I reached home, I thought about spending the rest of the afternoon with you and Mom. I called you but you have been driving, so I called mom and came here," said Jacob as he laughed.

"That's great, my son! Come in. We will cook together this afternoon," said Aidan.

"I would love to, Dad." With this, Jacob went inside the house. He hugged his mother and then all of them sat around the table.

Jacob told them how he felt when he waited for the managers to decide if he was eligible for the loan or not.

"I am proud of you, my son," said Jacob's mother.

"I wanted Dad to come for my support and tell me I can take such an initiative for myself," said Jacob.

"Son, I should have known before, but even if I wouldn't have shown up, always trust your skills. You do not require anyone's validation. Remember that," replied Aidan.

"You know, Dad, I had sworn to myself that even after the show, if you wouldn't support me, I would try my best to get the loan and then convince you to help me. See, I have the courage to execute things but you have the experience and you know the grind, so I will always need your guidance."

"I will always help you, son. I always knew you could do it, but as I told you, I feared your suffering," replied Aidan.

"Don't worry, men. Everything will be fine!" added Jacob's mom.

All three of them laughed and then the father and son cooked the steak together. They enjoyed cooking together and spent quality time with each other.

"Chef Jacob and Chef Aidan, this is the best steak I have ever ate; I bless your endeavor," said Jacob's mom.

Both Jacob and Aidan hugged her and Jacob said under his breath, "Oh God, thank you for this moment."

***

Two days later, Jacob received a check from the bank worth $200,000. He got in touch with the contractors and the remodeling of the restaurant soon began. It was the same location where his father had his restaurant. It was like a new beginning, but with Jacob.

This time, he had everything—finances, his parents' support, and expertise. The construction stayed in full swing. Jacob had started massive marketing for the restaurant. People began to pre-book for the grand opening.

After two months' time, the restaurant opened. It was a houseful affair. Jacob greeted everyone, and he served free drinks to everyone who attended. People left thank-you notes for the restaurant management for being a good host.

As the opening ended, many people came to Jacob and congratulated him in person.

Seeing that, Aidan turned to him, hugged him, and said, "My son! I am proud of you!"

What Jacob taught us is that when there is a will, there is always a way. We shouldn't let what others say stop us. Our opinions may be different and they may not understand our vision, but when we know our potential, nothing should stop us.

12

Another Chance

After having heard the story by Uncle P, Amanda was moved, and so was Jeffrey. She didn't know what to say. She had been at the right place with the right man. She had spent a long time thinking it was her shortcomings that led to a disastrous divorce. Her ex-husband didn't even talk to her once about what had gone wrong, so she thought it was her mistake. She had been penalizing herself for so long now. It was high time, and she had understood life would not wait for her to make the correct decision. Life will move and it will change, and we have to keep up with it.

Amanda was afraid that she might be cheated on again, and she knew if she was broken into pieces again, she could not gather herself and start once again.

Jeffrey also didn't talk about this.

"This show was interesting. I liked it a lot. It was a meaningful show," said Jeffrey.

"Yes, it was an amazing show. For any wise person, it had a lot of lessons to learn," replied Amanda.

Both of them made their way to the parking lot and got in the car. Jeffrey began to drive.

The wind felt cool. "Please don't turn on the AC. I like the wind. Let's enjoy that on our way back," said Amanda.

"As you wish," said Jeffrey as he rolled down the car windows.

Amanda leaned toward the window and looked outside the car. She didn't talk much. She enjoyed the wind and smiled throughout the journey. She kept contemplating the story but didn't say a word.

Soon, they were outside Amanda's mansion. Before getting out of the car, Amanda said, "For so many years in my life, I have feared to explore my love. I always feared disappointment, and because of the bad experience of the pass, I closed the doors of happiness for myself." With this, she began to cry.

"Oh my dear Amanda, why are you crying?" asked Jeffrey with concern.

"After having been married with a man that didn't appreciate me and being betrayed by him, I am letting go of the man who truly loves me. Why am I doing this to myself? I never deserved a toxic man, nor do I deserve to let a good man go," cried Amanda.

Jeffrey held her hands. She looked at him, but she didn't say a word. Jeffrey too stayed silent. Both of them enjoyed this moment of silence and comfort around each other. None of them said even a single word. It was like; the time had come for both of them to let go of loneliness.

Amanda and Jeffrey had understood they were made for each other. The interview, the timing, and the show were all set by God. He wrote it in destiny. Fate had guided both of them toward each other. Prior to this, they both lived in different parts of the world, but even the geographical distance couldn't separate them. They were meant to be.

"Jeffrey, I like you a lot. I have always admired you. I would love to be with a man like you, but my fear—my fear comes over my fondness of you." Amanda finally broke the silence.

"Amanda, everyone in life needs to let go of the old furniture. Don't let the loneliness dwell. Don't let it take a toll on you; we all deserve to live a happy life. Don't let these fears cage you. We only get one life, so make the most of it," replied Jeffrey and then he kissed her hands.

Amanda blushed. Even at this age, she had a heart like a teenager who would be moved by this act of affection. Amanda then leaned forward and kissed Jeffrey on the cheek. Just like Amanda, Jeffrey also blushed.

"A very good night to you," said Amanda as she stepped out of the car. "For I have seen an angel today." Amanda smiled. She closed the door of the car and began to wave until Jeffrey drove away. The main gate was opened and Amanda went inside. She unlocked the main door of the mansion and went inside. She found Elise sleeping on the sofa. She woke her up.

"Elise, why didn't you go to bed? Why are you sleeping here?" asked Amanda.

"Ma'am, I was waiting for you. How was your day?" said Elise with her eyes half open. She, too, wanted to know if Amanda was fine with Jeffrey.

Amanda let out a sigh. She sat by Elise's side and said, "I know what you want to know and I have decided upon it. I will proceed with this new man in my life. I will give him and myself a chance."

"Oh ma'am, I am happy for you," said Elise as she hugged Amanda.

Amanda smiled and said, "I am happy to know you are happy for me, Elise. But now, go to your room. Your back will hurt if you spend more time on this sofa."

Elise laughed and said, "Yes ma'am," and then she went toward her room. So did Amanda.

The next day, Amanda called her kids. She told them she moved forward in her life and will give Jeffrey a chance. Her kids, who were happy, were also concerned if it was the right step or not. However, after a few months' time, Jeffrey proposed to Amanda.

Amanda gave herself a chance to live a life that she dreamed for and to be with someone who would love her unconditionally. Amanda and Jeffrey lived a fulfilled life.

We can understand from their story that it doesn't matter whether we belong to the same place or same walks of life—if our hearts are at peace with each other, it is the best achievement that people can have in a relationship.

We should never close the door of love for ourselves. How can one live a life without love? Why should one let go of someone who can give them all the happiness of the world? We all have one life, so why live it with regrets?

***

Emma, who attended the show with her kids, now understood that she would always have to step out of her comfort zone to achieve something big.

It is imperative that achieving our goals will require extra efforts and someone who wants to make a difference will do all that it takes to achieve that goal to make the difference.

Emma's kids, who previously dreaded going back to Afghanistan, now began to support their mom. Both understood that fear never allows us to grow, and fear will only stop us from achieving our goals. Fear makes one become indecisive, and we are always one decision away from a different life.

Emma's classmates who had promised to help her go back to Afghanistan began with the preparations.

As soon as Emma graduated from Ronald University as a social worker, the family would leave for Kabul, Afghanistan.

During the airplane journey, Emma and the kids felt nervous. Emma was so close to her dream come true. She was all set to help the women of Afghanistan and would see her family after a long time.

The plane touched the grounds of Kabul. Emma and the kids were finally home. Emma had a strange feeling in her gut. She didn't know if she was happy or sad. She was rather overwhelmed. She didn't know if she had just come home or if she had left her home.

After having picked up the luggage, Emma began to move toward the exit, but her kids didn't move. She kept the luggage on one side and asked, "Boys, what's up? Why aren't you guys moving?"

"Mom, we don't know if it will be good," said Ibrahim, her elder son.

"Why, my son? What makes you think like that?"

"Mom, the feeling is different. We have left behind a place that gave us shelter and security. This land is very familiar, yet so unknown," told her son.

"I understand your feelings, son, but why are you letting them hold you back? We are so close to pursing our dream. Coming back to our land is a new beginning for the three of us. Think about is as being reborn. It is up to you how you shape this experience, embrace what life brings forward, and be determined to make a difference. I am not saying this new life will be easy, but it will be something worth fighting for."

Upon hearing this, her sons nodded and smiled. Emma picked up the luggage, and they headed to the exit. The afghan winds hit the family. Emma took a deep breath. Emma didn't know if she could go to her family or not. In reality, she didn't want to go there.

Suddenly, two security agents arrived and asked, "Are you Emma Dil?"

"Yes," replied Emma.

The kids became frightened. How can someone at the airport know about her presence?

"Come with us; we are from the American Embassy." One officer showed his card.

"The American Embassy had informed us of your arrival. There is a vehicle waiting for you and Mansur Kadal has sent you the key to a house that has been donated and furnished by his family as a gesture of honor for you. We will take your suitcases."

With this, the officers took her suitcases and told Emma to follow them. Emma clutched her sons' hands tight and followed the officers. It was like the three of them had been granted a new life.

Emma looked up to the sky and said, "Oh, you who takes care of all of us, never leave us, what will we do without you?"

Emma knew coming back to Afghanistan had embarked them to a whole new chapter of their lives. It would not be easy, but it would not be impossible.

After settling into her new home, Emma visited her ex-husband's grave with her kids.

Her kids began to weep as they sat by the grave. Emma sat by the grave and said, "I forgive you."

After sitting for a while next to his father's grave, Ibrahim got up and walked about fifteen feet from his father's grave. Emma, who didn't know what Ibrahim was up to, looked at her son. Ibrahim looked around the cemetery. He realized the cemetery where his father was buried was large. He thought for 30 seconds, then turned around and walked back to where his mother and younger brother Ali stood.

"Mother, after everything we have gone through and all these years of pain, I think there is a purpose for our lives. God has sent us for something," said Ibrahim.

"I think so too, my son ... nothing in this world is without reason," replied Emma.

"What made you think like that?" Ali added.

"The way our father died and this large cemetery. I would like to go back to the United States to study medicine and then return to my homeland. I see a great need for doctors in our country. Many people must have died because there wasn't enough medical help for them to survive. I want to help save lives. How long are we going to let people die like this?" said Ibrahim.

Emma had tears in her eyes.

Then he continued, "Mother, you can count on me, because our return here is to 'challenge the unknown with a purpose.'"

End Note

From the stories enclosed above, what did you learn as a reader? Don't we all have fears that invade our heart? If something is holding us from making decisions that can change our lives in its entirety, why are we trying to let the fear of the unknown take over?

It's the unknown we fear, not the situation itself. We do not know what the future holds and where our decisions will lead, causing us to fear.

We all need to understand one thing: we cannot let the fear of unknown win. Rather, we will have to learn the art of challenging the unknown.

Thank you for coming with us until the very end

Author's Commentary

This book is to motivate you to get out of your comfort zone and have some courage to challenge the unknown in your life. Fear will get you nowhere and you could miss out on great opportunities in your life. Staying in the same place fighting the same situation over and over again is a choice, not your destiny.

It is possible that your happiness is closer than you can imagine. Don't close the door to new opportunities in your life, nor let the negative voices of people who have achieved nothing in their lives discourage you.

Follow us on:
http://www.alfredophipps.com
https://twitter.com/alfredoephipps
https://www.facebook.com/alfredoephipps
https://www.instagram.com/alfredoephipps

CPSIA information can be obtained
at www.ICGtesting.com
Printed in the USA
BVHW030358111120
592859BV00028B/76